PLANT TISSUE CULTURE: RELEVANCE TO PROPAGATION, CONSERVATION AND CROP IMPROVEMENT

PLANT TISSUE CULTURE: RELEVANCE TO PROPAGATION, CONSERVATION AND CROP IMPROVEMENT

Edited by

Dr. Esha Rami

Delve
PUBLISHING

www.delvepublishing.com

Plant Tissue Culture: Relevance to Propagation, Conservation and Crop Improvement

Dr. Esha Rami

Delve Publishing
224 Shoreacres Road
Burlington, ON L7L 2H2
Canada
www.delvepublishing.com
Email: orders@arclereducation.com

ISBN: 978-1-77469-803-7 (Hardcover)

ABOUT THE EDITOR

Dr. Esha Rami is presently working as an Assistant Professor, in the Department of Life science, Parul Institute of Applied Science, Parul University, India. She Did her Post graduated from Ganpat University, Ph.D. in biotechnology from Hemchandracharya North Gujarat University, in the year 2015. She has authored a number of national and international publications in reputed journals.

TABLE OF CONTENTS

LIST OF FIGURES

PREFACE

Plants are immobile and are hence subject to high levels of biotic and abiotic environmental stresses. They have defense systems to withstand stressful conditions due to biotic and abiotic stressors. Global climate change, pollution, population, pests, and other related problems have made the current environmental condition much worse, directly affecting food production. Therefore, it is crucial for the scientific community to comprehend the consequences and varied stress tolerance systems in plants. This book aims to combine some of the biochemical, physiological, and molecular aspects of plant tissue culture. The focus of tissue culture research during the past twenty years has been on its industrial and agricultural applications. Several issues were discovered throughout the process of converting laboratory protocols into commercial protocols, and research efforts were directed toward addressing these issues. Most of the research on somatic embryogenesis focused on the differentiation of structures that resembled embryos. However, when the protocols were rigorously examined for use in commercial plant propagation, it became clear that the somatic embryos exhibited a very low degree of germination because of their physiological and biochemical immaturity.

The economic value of plant cultures is encouraged in all parts of the world, and *in vitro* culture gained popularity not only where the cost was recovered by the added value of plants in the market but also in applied research where *in vitro* plants are there. Today, no biotech lab may disregard *in vitro* multiplication after mutant recovery, for pathogen-free plant production, for regenerating mutants, and for other purposes. Since ornamental plants are primarily grown for their aesthetic value, the ornamental business places a high priority on the propagation of high-quality traits such flower color and scent. Due to low labor costs, decorative plants now have a major economic value that is rising by 8–10 percent annually, especially in developing nations. The cost of labor is particularly high, driving up the price of propagules and leaving businesses with a little profit margin. Because ornamental plant micropropagation relies heavily on labor, plant multiplication tasks are increasingly being outsourced to nations with lower labor costs. Despite all of these issues, *in vitro* plant propagation is still a successful method that is utilized by numerous national and commercial labs across the world for secondary metabolite production, rapid plant multiplication, and germplasm conservation. Millions of crop plants are regularly generated *in vitro* each year. By using low-cost tissue culture, adopting procedures, and maximizing the use of equipment and resources, it is possible to harness the enormous potential of micropropagation for large-scale plant production. This will lower the cost of production per plant without sacrificing quality. To satisfy consumer demand throughout the year, new crop varieties must be created. The ideal strategy for creating new food producing cultivars would combine *in vitro* culture and mutagenesis. The industry and academic institutions would greatly benefit from the established and refined techniques for *in vitro* culture and transgenesis and their direct applications in enhancing and creating new cultivars.

This book provides information about plant tissue culture and it can be used as a guide for undergraduates, postgraduates and researchers with basic knowledge of biology.

Introduction

Agriculture is an important business all over the world with limited exceptions. It is obvious that food production has reduced relatively due to climate change and other related factors. The reduction in food yields has a direct effect on the growing population and this will lead to distress among the nations in immediate future. It is essential to design techniques that can cater to the food requirements with limited financial resources. Plant tissue culture is one of many techniques that is being used in many areas to address food demands. Plant tissue culture can be used across all areas of biotechnology and can aid in the production of commercial crops and food crops with ease. Several disease-resistant plants were produced using this technique. The establishment of a plant tissue culture facility is a challenging task as contamination is a serious issue. The designing styles of such a facility is continuously refining due to the influx of pathogens through diverse ways or media. It is always essential to keep updated with the latest methods or protocols that minimize the loss while producing plants through tissue culture.

The production of media is another crucial issue that is to be considered after the establishment of the plant tissue culture facility. MS medium supplemented with phytohormones and nutrients needs to be balanced to generate better plant yields. *Agrobacterium*-mediated genetic transformation is being used in plant tissue culture labs to get the desired genes into the transgenic plants. Several virus-free plants have been produced using shoot-tip culture and this process is not possible outside the plant tissue culture facility. The greenhouse facility is another important infrastructure that can be considered to grow the plantlets produced from tissue culture. The greenhouse needs continuous monitoring to reduce the influx of plant pathogens and this requires experienced professionals.

CHAPTER 1

Introduction to Plant Tissue Culture

CONTENTS

1.1 INTRODUCTION

Fertilization, which involves the joining of two gametes of different sexes or strains to create a zygote, is one of the most important biological processes in the life cycle of an organism (Sandhir et al., 1999). The complete multicellular and multi-organalle body of a higher organism, whether it is a flowering plant or a human body, develops from this single-celled zygote. For instance, in a flowering plant, the single-celled zygote undergoes millions of mitoses before developing into structures as morphologically and functionally diverse as subterranean roots, green photosynthesizing leaves, and flowers (Khanday & Sundaresan, 2021). The latter kind of cell division is distinguished by identical outcomes. Thus, theoretically, the genetic material that was original y present in the zygote should have been transferred to all the cells in a plant body, whether they were found in the flowers, tissues, or root tips. All of this would indicate that there must be more factors acting in addition to the genetic makeup of cells to cause the large diversity shown by the genetically identical cells. Differentiation (Figure 1.1) is the process that leads to the manifestation of these variants. There are cellular and subcellular changes that occur before morphological differentiation (Hoffmann et al., 2021).

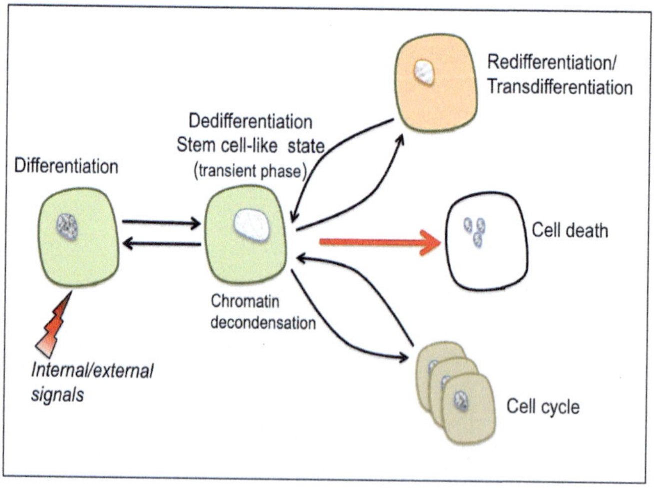

Figure 1.1. Cell differentiation.

(Source: Grafi, Gideon & Florentin, Assa & Ransbotyn, Vanessa & Morgenstern, Yakov. (2011). The Stem Cell State in Plant Development and in Response to Stress. Frontiers in plant science. 2. 53. 10.3389/fpls.2011.00053., Creative Commons License)

1.2 TISSUE CULTURE

Haberlandt (1902) selected single, isolated leaf cells for his tests. He employed Lamium purpureum and Eichhornia crassipes tissue, Ornithogalum epidermis, and Pulmonaria mollissima epidermal hairs (Kutschera & Ray, 2022). He raised them using Knop's (1865) salt solution (Camargo et al., 2019) with sucrose, and the palisade cells showed clear signs of growth. Since hundreds of thousands of plants can be produced from a tiny amount of tissue, differentiation of plants from callus cultures has been proposed as a potential technique for the quick replication of particular plant species. However, this approach has a significant flaw in that cells in prolonged cultures are genetically unstable. Ball created a more significant method in 1946, which would eventually grow into a practical horticulture method (Dagla, 2012). By cultivating the branch tips of Lupinus and Tropaeolum plants along with a few leaf primordia, he was able to successfully grow transplantable complete plants of both species. However, Morel and Martin were the first to successfully restore virus-free Dahlia plants from infected individuals by culturing shoot tips in vitro, demonstrating the practical applicability of this crucial approach (Bhojwani & Dantu, 2013b). The idea behind this method is that even in a plant that has been infected with a virus, the cells in the shoot tip are either virus-free or only slightly affected. Since then, shoot tip culture (Figure 1.2) has been widely utilized with a range of plant species of significant horticultural and agronomic relevance. It is now considered a normal procedure to produce virus-free plants from infected stocks.

Figure 1.2. Shoot-tip culture. (Source: Hu, G., Dong, Y., Zhang, Z., Fan, X., & Ren, F. (2022). Effect of In Vitro Culture of Long Shoot Tip on Variant Structure and Titer of Grapevine Viruses. *Plants*, *11*(15), 1907. https://doi.org/10.3390/plants11151907.

Morel (1960) used the shoot-tip culture technique to raise virus-free individuals of an orchid (Pant & Thapa, 2012). This approach might potentially be used for the quick propagation of these plants. Using this method, it was possible to grow 4 million genetically identical plants in a year from a single bud. Orchids were propagated up until this point using seeds. The progeny's seeming wide range of variation is an issue with this strategy. Commercial orchidologists quickly adopted this unique technique as their primary method for propagation after realizing how advantageous it was. In addition to revolutionizing the orchid industry, Morel's invention inspired the use of shoot-bud culture for the quick cloning of other plant species. Murashige played a key role in elevating in vitro culture techniques to the position of an effective practical method for horticulture species propagation. He contributed significantly to its widespread use by creating accepted techniques for the in vitro multiplication of numerous species, including ferns, leaf, flower, and fruit plants. It is based on another significant discovery made by Wickson and Thimann in 1958 (Sachs & Thimann, 1967). They demonstrated that the exogenous administration of cytokinins can start the growth of axillary buds, which are quiescent in the presence of terminal buds. This suggests that cultivating a developing stem with an undamaged terminal bud in a media containing cytokinin could cause the release of lateral buds (Gaspar et al., 1996). This would free the buds on the first segment of the stem as well as the lateral branches that sprouted from it from apical dominance in cultures, resulting in the development of a broom-like structure with several shoots. Frequently, a fraction of the total shoots may be rooted in a different medium to produce entire plantlets that are prepared for cautious handling and transfer to soil. An overview is given in Figure 1.3.

The type of project being done and the amount of funding available determine the size of a tissue culture setup and the degree to which it is equipped. The following facilities should be available in a typical tissue-culture lab. They are washing and storing glass, plastic, and other labware, preparing, sterilizing, and storing nutrient media, handling plant material aseptically, maintaining cultures under precise conditions of temperature, light, and humidity, observing cultures; and acclimating in vitro developed plants (Orlikowska et al., 2017). A media room for preparing media, washing glassware, and storing it should be available for research activity, and a growth/culture room should be provided to store cultures. A culture observation table and a sufficient light source should be present in the culture room. The sterile transfer cabinets may be kept in the culture room depending on the local conditions. A more complex setup is necessary for a commercial setup. The facility should ideally be shielded from any storms that may bring strong winds and rain, which might spread spores, mites, and thrips. To save energy, the growth chamber and the transfer room need to be sufficiently insulated. In certain instances, this has been accomplished by trapping air between two walls. The air between the two walls could be vented for a benefit during a hot season.

1.3 FACILITY REQUIREMENTS

A tissue culture facility needs plenty of high-quality water and a place to dispose of its waste(Abebaw et al., 2021). Where there are no public water or sewer facilities, this component needs special treatment. Local health and environmental municipal rules also apply to the disposal of any garbage. To avoid the sudden change in temperature in the culture room due to power outage and the shutdown of transfer hoods during the operation, which could occur even where a dependable source of electricity is available, a generator backup must be provided, at least to the culture room, and other essential equipment.

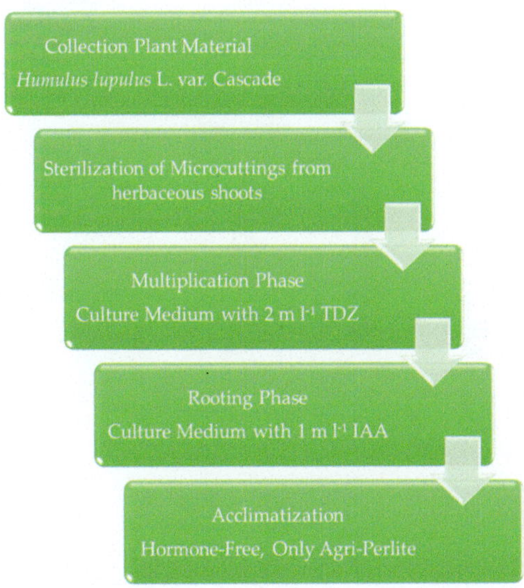

Figure 1.3. Tissue culture overview. (Source: Iacuzzi, N., Salamone, F., Farruggia, D., Tortorici, N., Vultaggio, L., & Tuttolomondo, T. (2023). Development of a New Micropropagation Protocol and Transfer of In Vitro Plants to In Vivo Conditions for Cascade Hop. *Plants*, *12*(15), 2877. https://doi.org/10.3390/plants12152877.

The shape and size of the available land as well as the anticipated capacity would determine the size and design of a facility (Chandra et al., 2010). To facilitating the frequent movement of materials between regions, a single-level construction that allows for simple access to diverse work areas is preferred. The design of the rooms, including the windows, doors, and hallways, must permit a workflow pattern that ensures the highest level of cleanliness and encourages the least amount of contamination (Cassells, 2012). It is important to demarcate the unclean (washing rooms, medium preparation rooms, autoclaving rooms, general stores, and offices) and clean (transfer and growth rooms) areas. Commercial laboratories should typically keep the air pressure positive, if not throughout the entire building, then at least in the clean section. These safety measures are essential to prevent the spread of contaminants into the culture vessels. It should be possible to maintain a higher level of asepsis in the clean area by moving materials into (sterilized medium, instruments, water, etc.) and out of (glassware and other things for washing and sterilizing, tissue culture-produced plants for hardening, etc.) the clean area through double door system (Rowe, 1986).

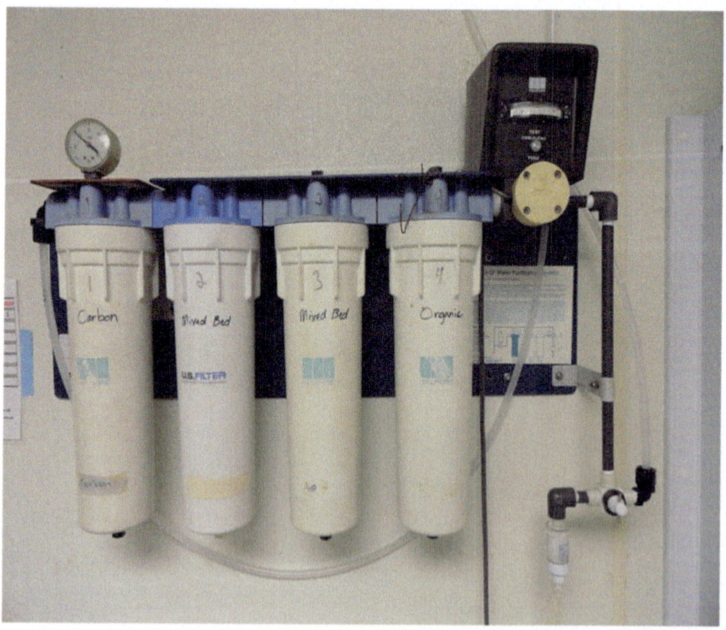

Figure 1.4. Milli-Q filtration system. (Source: https://upload.wikimedia.org/wikipedia/commons/thumb/0/0a/Milli-Q_Water_filtration_station.JPG/800px-Milli-Q_Water_filtration_station.JPG?20120928131223, Creative Commons License)

Brushes of all sizes and shapes, a washing machine, a sizable sink (ideally coated with lead to withstand acids and alkalis), and hot and cold running water should all be available in the media room. For the final washing of glassware and the preparation of media, a sufficient quantity of water is essential. Making provisions for water purification is necessary because tap water cannot be used to prepare a medium (Bhoite & Palshikar, 2014). For academic laboratories, deionized water may be utilized; however, water distillation equipment, a reverse osmosis unit, or water purification systems must be installed for research and commercial use. A glass distillation machine with sufficient handling capacity can be adequate for a research lab.

A Milli-Q filtration system (Figure 1.4) may be used in commercial establishments to give enough clean water that is free of organic pollutants, ionic contaminants, colloids, pyrogens, and traces of particles and microorganisms (Bozzo et al., 2002). To keep purified water, suitable storage tanks should be put in place. The typical equipment needed for culture medium preparation includes benches at a height that allows for standing work, a deep freeze for storing stock solutions, enzyme solutions, coconut milk, etc., and a refrigerator for storing various chemicals, plant materials, short-term stock solution storage, etc., plastic carboys made of plastic for holding distilled water, weighing scales, a hot plate-cum-magnetic stirrer for dissolving chemicals, a pH meter, an aspirator or vacuum pump to make it easier to sterilize filters and an autoclave or pressure cooker for sterilizing media (Bhojwani & Dantu, 2013a). A refrigerator may be kept near the media room in a hallway or another laboratory. Weighing balances shouldn't be used in the media room. An alternative would be to place a tiny weighing chamber in a dry area of the media room. Glass culture vials and other labware needed for media preparation have been substantially replaced with appropriate plasticware in many laboratories. Several plastics can be autoclaved. The market now offers a variety of disposable, pre-sterilized culture vials (made of clear plastic) that are specifically made for protoplast, cell, tissue, and organ culture procedures. Traditionally, cotton plugs have been utilized for culture tubes and flasks. However, there are many different alternatives. There are numerous metallic (aluminum and stainless steel) and plastic (polypropylene) caps available (Neumann et al., 2020).

Plant materials susceptible to gaseous buildup within the culture vials were tested in gas-permeable fluoro carbonate vessels (Lowe et al., 1998). The temperature in the space where cultures are incubated is kept under control. To keep the temperature at or near 25 ± 2°C, air conditioners and heaters attached to a temperature controller are typically employed. It is possible to employ specialized incubators with built-in fluorescent lights for treatments at higher or lower temperatures. These can be mounted anywhere, including outside the culture room, in a hallway, or another lab. However, care must be taken when left in the hallway to reduce the possibility of individuals tampering with the adjustment knobs. It might be able to maintain various growth conditions in several rooms in commercial businesses with many growth rooms. Given the importance of cleanliness in this environment, extra care should be taken to evade any direct contact with the exterior. The flooring and wall paint should be durable enough to survive frequent washing. In general, cultures are grown in diffuse light. Additionally, some consideration should be given to sustaining cultures in complete darkness and greater light intensities. Automatic timeclocks can be used to control the lights (fluorescent tubes) daily (Zaid et al., 2011).

To prevent the medium from drying out too quickly, provisions for increasing humidity should be provided if the humidity in the culture room drops below 50% (Afreen, 2008). When there is a lot of humidity, cotton plugs become damp and there is a higher likelihood that cultures may become contaminated. Specially built shelves for storing cultures should be included in the culture room. While some laboratories have shelves mounted on the side walls of the space, others have shelves attached to strategically positioned angular iron frames. Wheels could be added to the culture racks to make use of the available space. The shelves can be constructed from stiff wire mesh or glass. There is a unique set of fluorescent tubes for each shelf. A more consistent temperature is maintained around the cultures due to insulation between the lamps and the shelf above. The tissue culture lab must always be connected to a greenhouse, glass house, or plastic house to cultivate the mother plants and acclimate in vitro created plants. The level of sophistication of this facility will be determined by the funding. In general, the inoculum can be treated directly with a disinfectant if the explant is hard and large enough to be handled without difficulty. For instance, complete seeds, de-coated seeds, or stem portions, respectively, are surface sterilized in the cultivation of mature seeds, the mature endosperm of euphorbiaceous plants, or nodal explants. However, the conventional method is to surface sterilize the ovary or the ovule and dissect the explant under aseptic conditions so that the soft tissues of the inoculum are shielded from the toxic effects of the sterilizing agent (Kasperbauer & Collins, 1972). Similarly, shoot buds or flower buds are surface sterilized and the explant is removed aseptically when cultivating cultures of sensitive shoot apices or pollen grains. These explants often don't contain any microbiological contamination (Shahzad et al., 2017).

Figure 1.5. Contamination. (Source: Ali, M., Boonerjee, S., Islam, M.N., Saha, M.L., Hoque, M.I., & Sarker, R.H. (2018). Endogenous Bacterial Contamination of Plant Tissue Culture Materials: Identification and Control Strategy. *Plant Tissue Culture and Biotechnology*..)

1.4 CONTAMINATION

Several researchers have utilized antibiotics and antifungal substances to reduce explant contamination (Figure 1.5) (Danby et al., 1994). The bacteria that infect tissues are gram-negative, which are less sensitive to the routinely used antibiotics, hence indiscriminate use of antibiotics may not produce any useful outcomes. Fatty acid profile and nucleic acid investigations can all be used to precisely identify microorganisms (Meï et al., 2015). However, the traditional approach of utilizing liquid media or an enriched agar medium may be used if these procedures are thought to be expensive. In comparison to gentamicin, neomycin, and rifampicin, streptomycin was less phytotoxic to Mentha spp. and more effective against endophytic bacteria. However, studies on plant cultures have revealed that antibiotics limit roots, general growth, and multiplication. When used with mercuric chloride, antifungals such as binomyl have been demonstrated to diminish fungal infection. As long as the cabinet is turned on, all impurities such as hairs, salts, flakes, etc. are blasted away by the ultraclean air flow, maintaining a perfectly aseptic environment in the working area. The usage of a Bunsen burner or a spirit lamp is in no way hindered by air flow. Ordinary laboratories in temperate regions employ air-flow cabinets. To extend the life of the filters, it would be preferable to place the cabinet in a culture room with double doors in tropical and subtropical climates where air dust is particularly common. The hood should never be positioned in front of a regularly used entrance or window. Opening culture vials during the beginning of new cultures or subcultures of established cultures increases the likelihood of cultures becoming contaminated. As a result, the majority of commercial tissue culture organizations have a separate transfer room in a clean location, where the highest level of cleanliness and the least amount of air turbulence are maintained. People should be able to move around this area only in certain ways. The walls in this space ought to be smoothed out to prevent the accumulation of dust. Additionally, they may become extended with the explant or tissues and continue to sustain growth for many weeks without exhibiting any signs of a deficiency. Microelements play a crucial role as catalysts in numerous metabolic processes (Gautheret, 1955). Co and Ni are two substances that can prevent the synthesis of ethylene. Due to solubility issues, the availability of ions might occasionally become crucial. The interactions between Fe/ Mn or Zn make it difficult to interpret the status of Fe. While too much Fe or EDTA might inhibit Zn uptake, high Mn can result in Fe insufficiency (Larbi et al., 2002).

There isn't much solid proof that amino acids are important or necessary in plant tissue cultures. The amino acid glycine, which may even be inhibitory at larger concentrations, possibly offers no support in the persistent growth of tobacco callus. The plant cells may use amino acids directly or they may act as a source of nitrogen. But when an inorganic source is insufficient or used up, an organic source of nitrogen is preferred. To prevent tissue from turning black and to manage the oxidation of phenolics, cysteine has been added to the medium as an antioxidant. To maintain the healthy growth of tissues and organs, it is typically necessary to give one or more growth agents, such as auxins, cytokinins, and gibberellins, in addition to the nutrients. The need for these chemicals, however, differs greatly depending on the tissue and is thought to be influenced by endogenous levels. The amount of growth regulators needed is extremely small. There are numerous synthetic compounds with growth-regulating properties, each with a unique activity and species-specificity. When developing a tissue culture protocol for a new plant species, testing of multiple combinations, concentrations, and combinations of growth hormones are frequently necessary. Before sterilization, the pH of the medium is typically adjusted between 5.0 and 6.0. However, pH 7.0 is best for maize endosperm callus growth on a fresh weight basis, whereas pH 6.1 was best on a dry weight basis (Skirvin et al., 1986). In general, a pH higher than 6.0 produces a rather hard medium, whereas a pH lower than 5.0 prevents the agar from gelling to certain extent. The pH of the medium varies at different stages of preparation and culture; the pH of the medium once the gelling agent has been added dramatically decreases during autoclaving. Once plant tissue is placed on the medium, the pH of the medium changes even more. Regardless of the initial pH adjustment, the plant tissue and the medium interact to bring the pH to an equilibrium. The pH is also influenced by the ratio of NH4- to NO3- ions in the medium. When NH4 is largely absorbed, the medium becomes acidic as a result of the release of H ions, whereas ingestion of NO3- ions increases pH as a result of the release of OH- ions. The availability of different mineral ions in the medium and their absorption by the plant tissue are therefore influenced by these pH fluctuations (Leifert et al., 1995).

1.5 MEDIA COMPOSITION

The medium components are all broken down into four main groups: minerals, auxins, cytokinins, and organic nutrients (sucrose, amino acids, inositol, etc.). Utilizing commercially available dry powdered media that contains inorganic salts, vitamins, and amino acids is now the simplest way to prepare media (Satdive et al., 2007). Following the addition of sugar, agar, and any other needed supplements, the powder is dissolved in purified water, and the remaining pure water is used to make up the remaining volume. The medium is autoclaved and the pH is adjusted. There are now dry powder versions of some common plant tissue culture media and a variety of full mixtures for the micropropagation of certain plant species. For routine tasks like the micropropagation of plant species for which the composition of the needed medium is well established, powdered media may be helpful. The use of powdered media in these circumstances ought to save time. One approach is to weigh and mix the necessary amounts of the components individually before preparing the medium. To make a series of concentrated stock solutions, however, is a more practical and accepted technique. For researching cell metabolism and the impact of various chemicals on cellular responses, single-cell systems have been found to offer advantages over intact organs and whole plants by plant physiologists and plant biochemists. The fast administration and removal of many chemicals and radioactive substances is possible with the free-cell method (Thorpe, 2007). Through the application of microbial genetics to higher plants, crop enhancement is made possible through the cloning of individual cells. A practical option for the creation of a huge variety of commercially significant phytochemicals is large-scale in vitro culture of plant cells (Loyola-Vargas & Vázquez-Flota, 2006). The isolation of mesophyll cells is frequently accomplished by gently grinding the leaves, followed by cell cleaning via filtering and centrifugation. In contrast to the enzymatic technique, mechanical cell separation offers at least two key advantages: cells are not subjected to the potentially damaging effects of enzymes; and cells do not need to be plasmolyzed, which is frequently desired in physiological and biochemical research. The mechanical approach of cell separation is not always effective. Only in situations where the parenchymatous tissue was loosely structured and had few sites of cell-to-cell contact could mesophyll cells be successfully separated using this technique.

Because this method is practical and broadly applicable, single-cell systems are typically separated from cultivated tissues for use in basic and applied research. Auxins and cytokinins are typically utilized (Centeno et al., 1996). Cultures are started by simply placing freshly cut sections of surface-sterilized plant organs on a nutrition media. The explant displays callusing on such a medium, which typically begins at the cut ends and subsequently spreads to cover the full surface of the tissue. To develop a sufficient amount of tissue, the callus is detached from the parent explant and transferred to a new medium with the same composition. The friability of the tissue may also be improved by repeated subculture on the agar medium, which is very desirable for growing a fine cell suspension in a liquid medium (Teixeira et al., 1995).

The growth of a single peanut mesophyll cell in a liquid media can be observed using time-lapse photomicrography (Figure 1.6). After 3-5 days in culture, we can notice that the leaf cell had grown in size to the point that it no longer resembled a palisade cell. Before true cell division, plastids accumulate around the nucleus (systrophy). The spongy parenchyma cells die while only the palisade cells undergo division. Peanut spongy cells are also capable of proliferating if the cells had been isolated adequately. Spongy cell division, however, happens a little later than palisade cell division. In order to evenly disperse the cells and fix them in a thin layer of the medium that is approximately one millimeter thick after it has cooled and solidified, equal quantities of the media are mixed and quickly spread out in Petri dishes. The parafilm is used to seal the dishes. Suspension cultures that contain both free cells and cell aggregates ought to be filtered using a sieve that lets only individual cells and small cell aggregates pass through (Kieran et al., 1997). Only the fine suspension is plated after the big cell aggregates have been discarded.

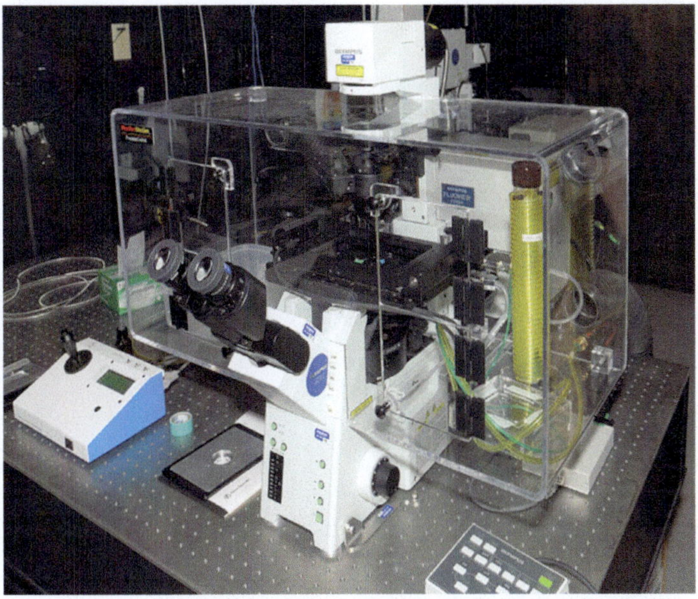

Figure 1.6. Time-lapse photomicrography. (Source:https://upload.wikimedia.org/wikipedia/commons/4/49/Olympus_FluoView_FV1000_Confocal_Microscope_-_NCMIR.jpg, Creative Commons License)

To assure the isolation of pure single cell clones, the plates can be seen under an inverted microscope and single cells can be marked with a fine marker on the outside of the plate. It appears that the callus gives the cell something else that is essential for cell division in addition to the nutrients from the culture media. The filter paper can allow the cell division factors to diffuse through it. By placing callus on an agar media, it is also possible to show how callus tissue stimulates the division of isolated cells. Cells near the calli begin to divide earliest in these cultures (Figure 1.7).

The theory that cells manufacture specific substances required for cell division has been used to explain the effect of cell density on cell division. Before a cell can begin to divide, the endogenous concentration of these substances must reach a certain level (Aremu et al., 2012). Up until an equilibrium between the cell and the media is attained, the cells continue to and release their metabolites into the medium. As a result, the equilibrium is reached much more quickly at high cell density than it is at low density, and as a result, the lag phase is longer in the latter case. The equilibrium is never attained below a threshold cell density, and cells are unable to divide. However, divisions can be supported at a relatively low cell density in a conditioned medium that is rich in the required metabolites.

Figure 1.7. Callus. (Source:https://d3i71xaburhd42.cloudfront.net/f8578164b6997483d3f994b3a111f54d9ede15e0/4-Figure2-1.png, Creative Commons License)

We don't know enough about the precise factors causing cell division, which accounts for our inability to culture individual cells in an entirely synthetic media. A thorough examination of the conditioned medium ought to provide some hints in this area. It has been suggested that mass-producing plant cells in vitro is a practical option to produce a wide range of high-value, low-volume phytochemicals. The design of bioreactors for plant cell culture has therefore received a lot of attention over the past two decades. In a glass or steel container known as a bioreactor, organisms are cultured. Bioreactors (Figure 1.8) can be equipped with probes to measure the pH, temperature, and dissolved oxygen in the culture and also have options for sampling the cultures, adding fresh medium, adjusting pH, mixing the cultures, and regulating temperature without jeopardizing the aseptic properties. As a result, it is possible to control and monitor culture conditions more closely than is possible with shaking cultures (Weathers et al., 2010).

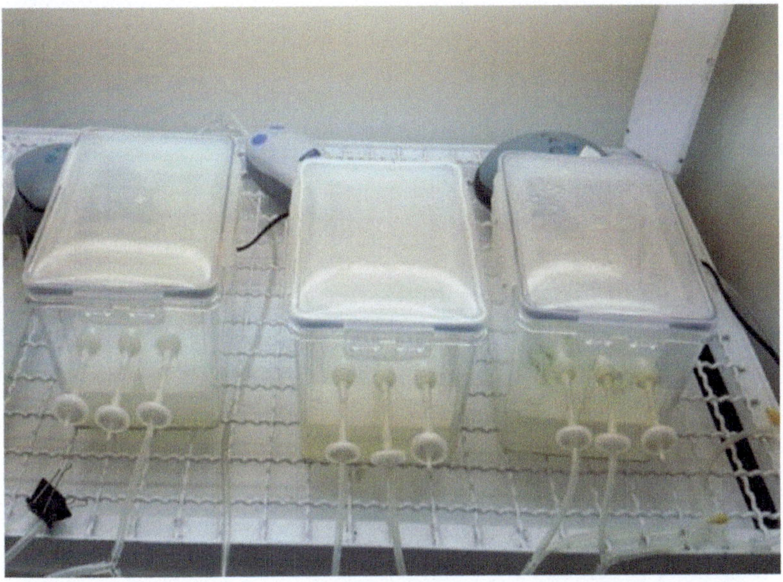

Figure 1.8. Bioreactor: PlantForm bioreactor TIS model consisting of one container. (Source: De Carlo A, Tarraf W, Lambardi M, Benelli C. Temporary Immersion System for Production of Biomass and Bioactive Compounds from Medicinal Plants. *Agronomy*. 2021; 11(12):2414. https://doi.org/10.3390/agronomy11122414, Creative Commons License)

1.6 FERMENTERS

The fermenters used for microbial cell cultures are not appropriate for plant cell cultures due to the differences in the nature and growth pattern of the types of cells, although the fundamental requirements for suspension cultures are similar to those of submerged microbial cultures (Leifert & Cassells, 2001).

Large-scale plant cell culture requires efficient mixing to maintain consistent physiological conditions inside the culture vessel. Despite having a higher tensile strength than microbial cells, plant cells are vulnerable to shear stress due to their size, tough cellulose wall, and numerous vacuoles, which prevents the use of strong agitation for effective mixing. As a result, plant cells are frequently cultivated at extremely low agitation rates in modified stirred-tank bioreactors. At low shear, air-lift reactors might offer even more consistent and beneficial environmental conditions. Since every plant cell is aerobic, it needs a constant supply of oxygen. However, due to their slower metabolism, plant cells need less oxygen than microorganisms. High oxygen concentrations can occasionally be harmful to metabolic processes (Jay et al., 1992). Normally, the air is blown in or sparged at the bioreactor's base. In suspension culture, plant cells frequently group in aggregates of 2–200 cells (Pépin et al., 1999). Cells become stickier in the late exponential phase of growth as a result of increased polysaccharide excretion into the culture vessel. As a result, plant cells stick to the reactor wall, probes, and stirring mechanism, forming bigger aggregates. Because the aggregates tend to settle out or attach to the reactor surface, where they create thick walls, mixing is hampered. Large aggregates can obstruct the reactor's opening and pipelines, which causes rheological issues by forming dead zones in the culture vessel.

Figure 1.9. Fermenters: Comparison of culture systems of taro (*Colocasia esculenta*) shoot multiplication after 30 days of culture. (a) Semi-solid medium, (b) temporary immersion bioreactor, (c) ebb-and-flow bioreactor (d) SETIS™ bioreactor. Black and white bar = 5 cm. (Source: Mancilla-Álvarez E, Pérez-Sato JA, Núñez-Pastrana R, Spinoso-Castillo JL, Bello-Bello JJ. Comparison of Different Semi-Automated Bioreactors for *In Vitro* Propagation of Taro (*Colocasia esculenta* L. Schott). *Plants*. 2021; 10(5):1010. https://doi.org/10.3390/plants10051010, Creative Commons License)

Cell aggregation compromises the performance of the probes used to track culture conditions throughout production (Panda et al., 1989). When nutrients can no longer reach the central core of massive aggregates, diffusion-limited metabolic events may take place. Despite these negative consequences, it appears that specific levels of cell aggregation and cell differentiation are necessary for the formation of secondary metabolites. Therefore, from the perspective of process engineering, regulated plant cell aggregation is of interest. In contrast to animals, where differentiation is typically irreversible, even highly developed and differentiated cells in plants still can regress to a meristematic condition if their membrane systems and nuclei are intact. The degree of regression a cell can experience will depend on the cytological and physiological state it has attained in situ. The cells first go through specific alterations to acquire the meristematic stage before being cultivated on a nutritional medium that supports their proliferation from non-dividing, quiescent cells from differentiated tissues. These include the replacement of damage by lysosomal activity during the phases of cytoquiescence non-functional cellular components. Dedifferentiation is the process by which a mature cell reverts to its meristematic state and produces undifferentiated callus tissue (Sugiyama, 2015). A multicellular explant typically consists of different types of cells. Because of this, the callus formed from it would be heterogeneous in terms of its cell capacity to develop into an entire plant or specific plant organs. Cellular totipotency refers to the innate capacity to develop into an entire plant, a capacity that is frequently retained even after a cell has completed final differentiation in the plant body (Vasil & Vasil, 1972). A differentiated cell must first go through dedifferentiation and then redifferentiation in order to express its totipotency. Dedifferentiation mostly results in embryonisation of cells, which causes callus development. But embryonic explants frequently show differentiation of roots, branches, or embryos without a callus phase in between (Carman, 1990).

1.7 PHYTOHORMONES

In addition to providing a chance to investigate the mechanisms that cause cells to be totipotent, tissue culture techniques also permit the exploration of the elements governing cytological and histological differentiation. The totipotency of somatic cells has long been utilized in the vegetative growth of several plant species (Farbos et al., 1997). Either shoot-bud differentiation or somatic embryogenesis (Figure 1.10) can result in whole-plant regeneration from cultivated cells. Based on physical morphological distinctions, a shoot bud and an embryo can be distinguished. Plant regeneration is typically more challenging from single cells, protoplasts, or unstructured calli than it is from whole explants like cotyledons, hypocotyl segments, and premature embryos (Norstog, 1973). With the development of methods to introduce foreign genes into the cells of intact explants, the difficult process of plant regeneration from isolated protoplasts is no longer necessary for the success of plant genetic engineering. Optimizing procedures for in vitro differentiation directly from immature embryos and seedling explants has received a lot of interest during the past ten years. Somatic embryogenesis is the focus of the majority of recent reports on in vitro plant regeneration because it may be more useful for plant propagation than organogenesis and is proving to be the best system for examining the cellular basis of differentiation in higher plants. When transplanted from a medium containing 2,4-D to a medium not having it or having NAA/IAA in its place, callus tissue from most cereals displays organogenesis.

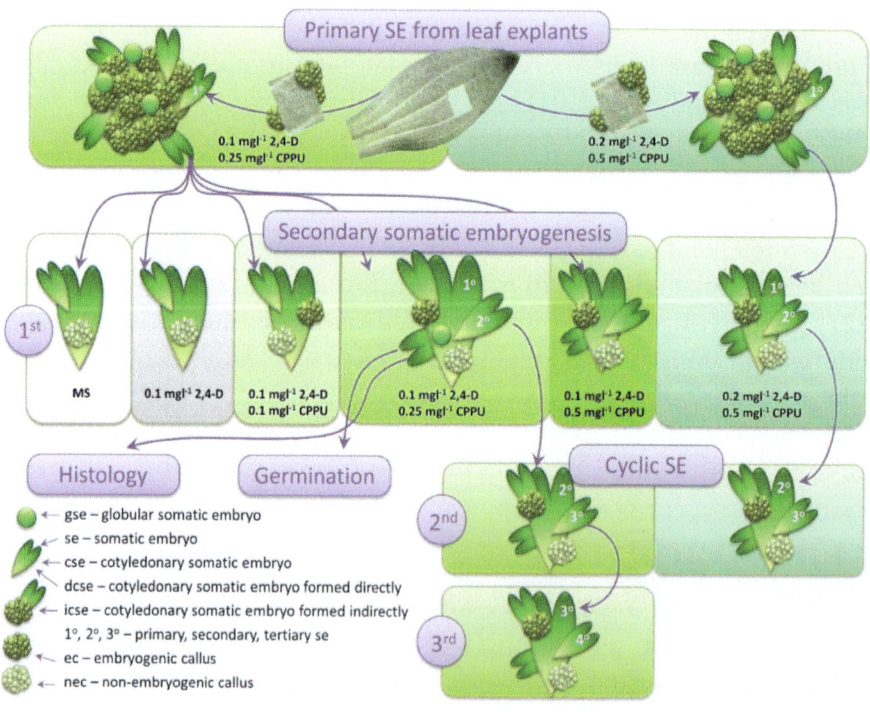

Figure 1.10. Somatic embryogenesis.

(Source: Bogdanović, M. D., Ćuković, K. B., Subotić, A. R., Dragićević, M. B., Simonović, A. D., Filipović, B. K., & Todorović, S. I. (2021). Secondary Somatic Embryogenesis in Centaurium erythraea Rafn. *Plants*, *10*(2), 199. https://doi.org/10.3390/plants10020199, Creative Commons License)

SE starts with asymmetric divisions of single totipotent cells from the epidermal or subepidermal layers of the root explant. Successive divisions give rise to somatic embryos. A shoot-bud distinction is less common in plants than in the production of roots. Alfalfa likewise undergoes a two-step process of organogenic differentiation. On a medium containing 2,4-D and kinetin, callus is first initiated and then proliferated (Li et al., 2012). When tissue fragments from such calli are added to a hormone-free solution, organogenesis occurs. The nature of the organ developed depends critically on the hormone ratio in the induction medium. Shoot bud differentiation was inhibited when the differentiating callus was exposed to GA3 in the dark for as little as 30 to 60 minutes (Mukhopadhyay & Bhojwani, 1978). After 48 hours, neither meristemoids nor shoot buds were still present. Gibberellin was particularly useful during the meristemoid development phase. Once shoot buds had formed, GA3 did not prevent their continued growth (Yamaguchi, 2008).

Gibberellin-like compounds are present in tobacco tissue, and the tissue can metabolize exogenous gibberellin. Gibberellin may play a role in the development of tobacco buds. The tissue most likely produces the hormone in proportions ideal for the organogenic process, which accounts for the inhibition caused by exogenous gibberellin. The enhancement of bud development by abscisic acid in sweet potatoes, however, can be explained by the presence of optimal levels of endogenous gibberellins. The suppression of bud formation caused by GA3 in tobacco is largely dominated by abscisic acid at a non-toxic concentration (Tanaka-Ueguchi et al., 1998). Significant amounts of ethylene and carbon dioxide were produced by the explants that formed shoots, and it was possible to link the frequency of shoot bud formation to the concentration of the gases inside the culture vial. When these gases were taken out of the culture medium, organogenesis was completely halted. However, high gas buildup led to partial dedifferentiation of the shoot buds. In addition to auxin and cytokinin, adding raw Panax quadrupled the regeneration frequency in cultures of cotyledons and hypocotyl of Brassica oleracea. Weak electric current application in tissue cultures can significantly improve organogenic and embryogenic differentiation. The application of a small electric current in such a way as to make the callus negatively charged and the medium positively charged resulted in a 70% increase in the formation of tobacco calluses (J. Y. Kim et al., 2019). This only took place on a medium containing IAA. Following the electric treatment, some shoot bud differentiation occurred on the callusing media, which typically did not favor any callogenesis.

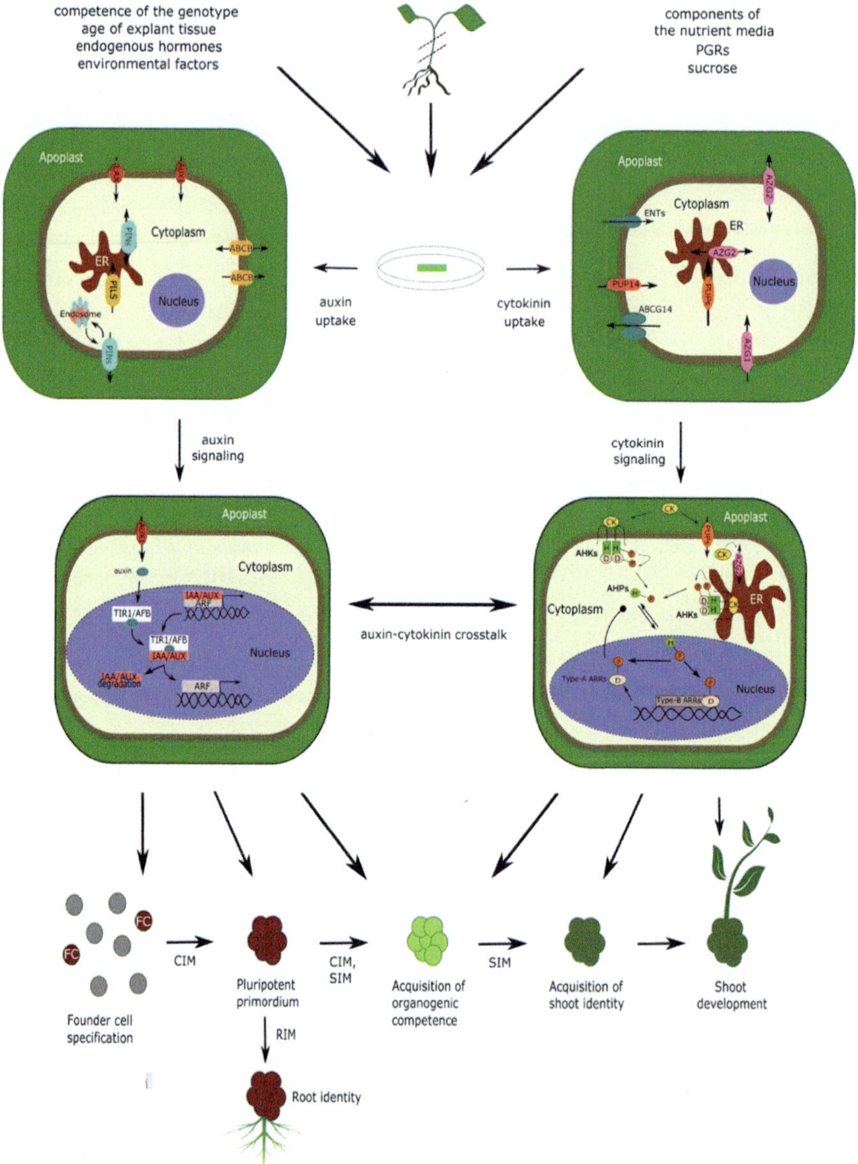

Figure 1.11. Organogenic differentiation. (Source: Raspor M, Motyka V, Kaleri AR, Ninković S, Tubić L, Cingel A, Ćosić T. Integrating the Roles for Cytokinin and Auxin in De Novo Shoot Organogenesis: From Hormone Uptake to Signaling Outputs. *International Journal of Molecular Sciences*. 2021; 22(16):8554. https://doi.org/10.3390/ijms22168554, Creative Commons License)

When microampere current was applied to the callus on a medium for shoot differentiation, the stimulation of shoot bud differentiation increased five times more than it did in the control (Rathore & Goldsworthy, 1985). Regardless of the polarity of the current, the shoot buds first formed in the callus- negative area. The number of cultures generating shoot buds and the number of buds per culture were both impacted by the electric stimulation of callogenesis. The callus created from mature wheat embryos, which normally only produced roots, was exposed to electrical treatments to cause it to produce numerous shoots. The age of the donor plant has an impact on the physiological status of an explant, which directly affects the explant's ability to regenerate. In several instances, raising regeneration cultures has been made possible by the use of immature, meristematic tissues when mature, differentiated explants failed to respond. For cereals and certain types of trees, this is especially true. The percentage of tissue cultures and the number of shoots per culture of Psophocarpus tetragonolobus seedlings both decreased as seedling age increased (Singh et al., 2014). Most regenerating cotyledons in Brassica juncea were seen in seedlings that were 5 days old. Explant preparation is also crucial. Only the cut end of the petiole forms shoots buds or roots in B. juncea cotyledon cultures, depending on the culture medium. This potential is absent in the lamina. However, for the petiole cells to demonstrate totipotency, laminar tissue must be present. In cotyledon, cultures of B. juncea, the orientation of the explant on the medium was found to be a crucial element for organogenic development. Explants that failed to produce roots or shoots within three to five days of culture were those in which the lamina expanded and curled, causing the petiole to lose touch with the medium.

Organogenic differentiation (Figure 1.11) in cell and tissue cultures is a multi-step process that happens in response to hormonal manipulation of the culture medium (Walker et al., 1979). Before the emergence of a morphologically zrecognizable organ, a sequence of intracellular activities known as induction takes place. Even if the inducing hormones are taken away, the cells that were stimulated to develop into a particular organ while being exposed to the right hormonal mixture would continue to do so. Similar to this, cotyledons pre-cultured on BAP-free medium for longer than 7 days lose their ability to develop into shoots on BAP-medium. Fertilization causes the egg cell to divide and develop into an embryo (Sihachakr & Ducreux, 1987). The egg cell is then referred to as the zygote after fertilization. However, the egg can sometimes be stimulated to go through embryogenesis without fertilization. The pollination stimulus alone or just the administration of certain growth regulators, as occurs in parthenogenesis, may cause the egg to undergo embryonic development. Furthermore, the ability to develop an embryo is not unique to the egg. An embryo can develop from any cell within the female gametophyte or even from the sporophytic tissue surrounding the embryo sac. It is common for adventive embryos to arise from nucellar cells in numerous species of Citrus and Mangifera (Lakshmanan & Ambegaokar, 1984).

There has been a surge in the species that produce somatic embryos (SEs) during the past three decades and a great deal of information has been gathered to determine the embryogenic potential of somatic plant cells (Fehér, 2006). It would be rational to assume that any cell whose irreversible differentiation has not advanced too far will, if placed in a suitable medium, develop in an embryo-like manner and generate a whole plant based on the recent remarkable development in the tissue culture of higher plants. Therefore, the entire complicated sexual apparatus is not a necessary condition for cells to develop embryonic traits. Endogenous hormone levels may be the cause of genetic differences. The embryogenic response increased tenfold after hormone removal. According to several studies, cytoplasmic factors may also be involved in the regulation of somatic embryogenesis. In vitro embryogenesis is substantially influenced by the type of nitrogen present in the medium. Embryo development only took place in cultures of wild carrots produced from petiolar segments provided the media contained slightly decreased nitrogen. After the auxin was removed, the calli that had been begun on a medium containing KNO3 as the only supply of nitrogen failed to develop into embryos. Polyamines are necessary for embryo development (Kakkar et al., 2000).

Mango cultivars with polyembryonate nucelli had much larger concentrations of putrescine, spermidine, and spermine than those with mono-embryonate nucelli. Increases in endogenous levels of polyamines, enzymes for their biosynthesis, and somatic embryogenesis in carrots, along with the suppression of somatic embryogenesis by inhibitors of polyamine biosynthesis, all occur concurrently with the induction of somatic embryogenesis. Celery globular embryos had a 37-fold higher polyamine concentration than plantlets; putrescine, spermidine, and spermine were the three polyamines examined, and putrescine showed the most pronounced increase (Scoccianti et al., 2016). Similar results were shown in mango embryogenic cultures, where putrescine levels significantly increased. However, the increase in mono-embryonate cultures was almost twice as great as the growth in polyembryonate variety cultures.

The electric stimulus appears to encourage the differentiation of organized structures (shoots/embryos) by altering cell polarity through changes in microtubule organization. The treated alfalfa protoplasts had more microtubules, which were thinner and more branched than those of the untreated control cells, two days after growth. The microtubules were arranged in parallel strands in the untreated cells, whereas they appeared as a network in the treated cells. Another notable result of the electric treatment was the induction of a division that was asymmetric and followed by a brief phase of cell growth, leading to spherical structures made up of a majority of small, irregularly shaped cells. The untreated cells, on the other hand, went through a symmetrical first division, and this, as well as future divisions, resulted in colonies made up of cells of identical size. Explants with several cells typically have a wide range of morphogenic potential among their cells. Under specific cultivation conditions, only a small percentage of these cells can exhibit their cellular totipotency. As a result, the calli produced from such explants are similarly diverse.

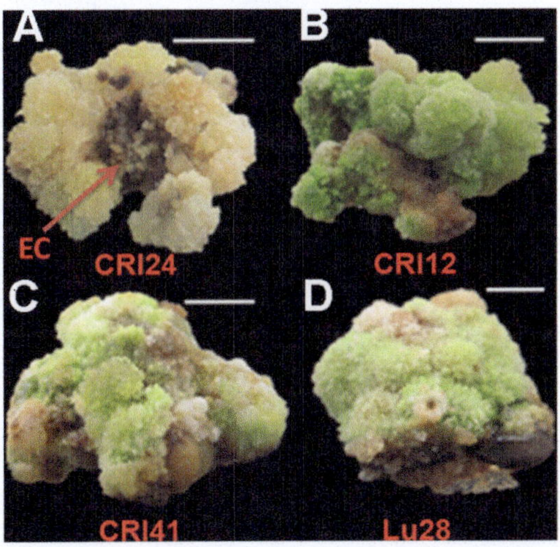

Figure 1.12. *Gossypium hirsutum* callus. A-D: callus type of four cultivars CRI24, CRI12, CRI41 and Lu28 (Source: Zheng W, Zhang X, Yang Z, Wu J, Li F, Duan L, et al. (2014) *AtWuschel* Promotes Formation of the Embryogenic Callus in *Gossypium hirsutum*. PLoS ONE 9(1): e87502. https://doi.org/10.1371/journal.pone.0087502, Creative Commons License)

Regardless of the explant utilized, two types of calli are generated in cereals: soft, granular, and translucent, and white or pale yellow, compact, and frequently nodular. Only the first type of calli among these displays embryogenic differentiation. Gossypium hirsutum callus (Figure 1.12) was categorized as green, yellow, white, brown, and red. Only yellow calluses produced cultures from embryos (Davidonis & Hamilton, 1983). Morphologically different embryogenic tissue must be subcultured at many periods to maintain conifer embryogenic cultures. Selective or artistic subculture has also been used to create embryogenic cultures of Daucus carota, Zea mays, Sorghum bicolor, Hordeum vulgare, and Picea abies. To differentiate somatic embryos, cells must be reactivated to divide. Depending on the type of explant and the culture conditions, embryos may form immediately from the explant cells or after several cycles of cell division. An unorganized growing period comes before embryogenic differentiation.

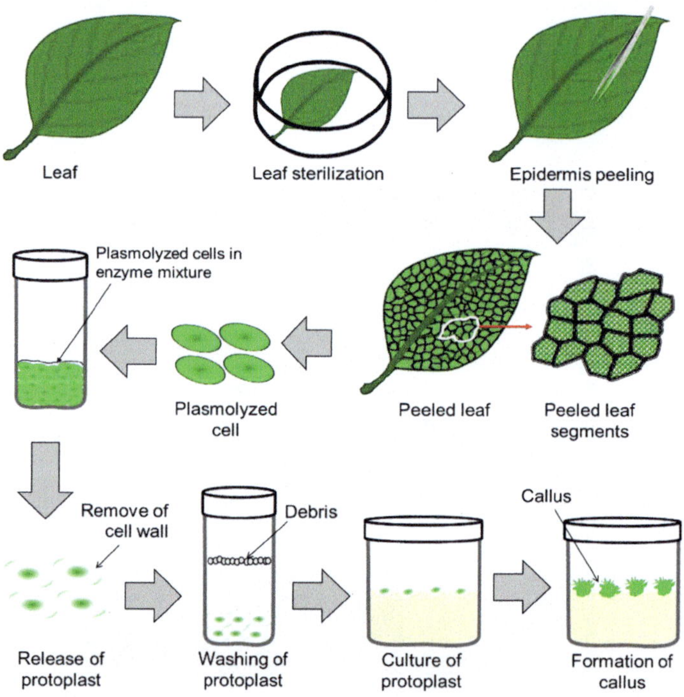

Figure 1.13. Protoplast cultures-an example protocol. (Source: Mehbub, H., Akter, A., Akter, Mst. A., Mandal, M. S. H., Hoque, Md. A., Tuleja, M., & Mehraj, H. (2022). Tissue Culture in Ornamentals: Cultivation Factors, Propagation Techniques, and Its Application. *Plants*, *11*(23), 3208. https://doi.org/10.3390/plants11233208, Creative Commons License)

Usually, direct embryogenesis takes place in cultures of immature embryos. Explants that can directly develop embryos appear to contain competent or preembryogenic-determined cells (PEDCs)(Chen et al., 1998).

In other instances, exposure to appropriate therapies results in the acquisition of embryogenic competence. The induced embryogenic determined cells or IEDCs are functionally equivalent to PEDCs once they have been induced. Both can survive and grow in the embryonic stage with the right environment. The idea that plant growth regulators and other therapies used to induce embryogenesis to accomplish this by changing cell polarity and encouraging subsequent asymmetric division is supported by some studies. White clover immature embryos grow numerous adventive embryos directly from the hypocotyl epidermis on cytokinin-rich media. A shift from the typical anticlinal divisions in the epidermis to periclinal or oblique divisions is the first cytological indicator of the induction of embryogenic cells. In protoplast cultures of an embryogenic cultivar of alfalfa, asymmetric cell divisions were commonly seen after stimulation by auxin; the protoplasts from a non-embryogenic cultivar divided symmetrically. An example of a protoplast culture scheme is shown below in Figure 1.13.

Only those embryos that remain linked to the callus until a suitable shoot meristem forms mature into whole embryos. They do not develop a normal shoot meristem when grown from cell clusters in a liquid medium. The globular embryos of the carrot, which readily develop full embryos in a liquid media, show electric polarity in the form of an ionic current; there is an influx of potassium ions at the eventual plumular end and an efflux of hydrogen ions at the radicular end (Ramos et al., 2015). Before the appearance of morphological polarity, the PEMs of carrots exhibit polarity in the distribution of the calcium-calmodulin complex. In the vicinity of the root pole, there was a greater concentration of activated calmodulin. Another way that cells might become polarized is by adhering to a substrate. By attaching PEMs to alginate beads to maintain a physiological polarity, the generation of embryos in liquid cultures of sweet potato was improved. The molecular mechanism that governs the process is still poorly understood. It has long been known that when somatic cells begin embryogenic development, changes in gene expression take place.

The presence of tiny, dense cytoplasmic cells in carrot cultures is a key factor in determining their embryogenic potential. When the SEs are inhibited by the glycosylation inhibitor tunicamycin in the globular stage, this results in cell elongation, which is a defining property of non-embryogenic cultures. In the presence of tunicamycin, a 38 kDa peroxidase isolated from embryogenic cultures suppressed cell elongation. As long as its enzymatic properties were unaltered, a commercial formulation of horseradish peroxidase assisted in normal embryo development when the treatment was present. The role of peroxidases is to maintain the size of the cells of early proembryos since tunicamycin induces the cells of very immature SEs to expand. Unlike seed embryos, SEs typically do not go through the final stage of embryogenesis, known as embryo maturation, which is demarcated by the accumulation of reserve food components and proteins that give the embryo desiccation tolerance. During this phase, the size of the embryo does not increase. According to reports, ABA, inhibits secondary embryogenesis and pluricotyledony and supports the normal development of embryos by preventing premature germination to increase embryo maturation in many species. Many different variables, including high-density inoculum, nutrient restriction, osmotic stress, and temperature shock, can replace ABA by causing the embryos to synthesize the hormone.

CHAPTER 2

Plant Propagation

CONTENTS

2.1 INTRODUCTION

In vitro procedures for the germination of orchid seeds were created by pioneers in plant tissue culture at the beginning of the twentieth century (Utami & Hariyanto, 2019). Later, a medium was created that allowed roots to grow as they floated on top of it. Tobacco vascular tissue extracts, malt extract, liquid endosperm, and a solid coconut endosperm extract were all shown to promote cell proliferation. Additionally, *in vitro*-cultured tobacco callus tissue induced cell division in response to autoclaved DNA from herring sperm and calf thymus. If the DNA had not been autoclaved, this activity would not have occurred. A group of growth regulators known as cytokinins promote cell division. It was discovered that the first cytokinin, kinetin, which was produced from animal DNA, stimulates cell division and shoot multiplication. Since then, new chemicals like benzyl adenine and thidiazuron have been produced. These substances are being used extensively in the commercial micropropagation sector and are substantially more potent than kinetin. It is now possible to develop somatic embryos, adventitious shoots, axillary shoots, and adventitious roots utilizing *in vitro* methods by adjusting the plant growth regulators, medium formulation, and plant materials. By the end of 1950, researchers started fusing tissue culture and thermotherapy (Bairu et al., 2011). Virus-infected plants (Figure 2.1) were cultivated at temperatures ranging from 25°C to 40°C. Plant pathologists have aided plant propagators by creating techniques for sterilizing medium, and planting containers in the 20th century (Sowley et al., 2010).

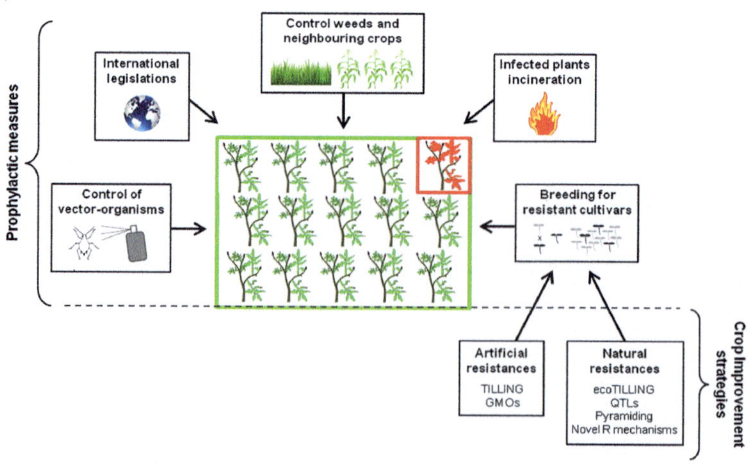

Figure 2.1. Virus-infected plants. (Source: Nicaise V (2014) Crop immunity against viruses: outcomes and future challenges. Front. Plant Sci. 5:660. doi: 10.3389/fpls.2014.00660, Creative Commons License)

2.2 PROPAGATION

The fruit, nut, and vegetable businesses of the twenty-first century depend heavily on viruses free plants (Wang & Hu, 1980). Successful seed germination is essential for large-scale agriculture of crops, including those for grains and some vegetables. Alternatively, one may argue that seeds constitute the first essential step in the global food chain and that this dependence has given rise to a sizable industry. The conditions necessary for germination might be as basic as the right combination of moisture and temperature. Large quantities of seed can sometimes be mechanically planted, but in some situations, they must be planted by hand to minimize damage, such as for buckeye, the seed coat with its barrier to water penetration must be broken or weakened (Quintero & Bowers, 2011). The horticulture sectors with the fastest growth rates are the flower and nursery industries. The most important contributor to this growth over the past 20 years has likely been the increase in plant demand and growth. Successful and uniform seed germination is necessary for the establishment of cultured plants. Even the forest business is highly interested in the methods of plant multiplication being utilized to enhance the performance of the trees, from the creation of seed orchards with superior tree species to the utilization of cuttings from superior clones. Somatic embryogenesis, or producing an embryo that is identical to the original from the somatic tissue of the plant, has been the subject of some efforts. For species like eucalyptus and pine, clonal methods are being studied all around the world, and molecular genetics methods are being investigated to augment the genes for auxin production in poplar to speed up how quickly the transgenic tree would grow.

The art and science of plant propagation are especially crucial for maintaining rare and endangered plant species. Species that are considered uncommon or endangered are propagated using both conventional and specialized propagation methods to protect them and, in certain circumstances, reintroduce them into both new and old environments. For species that respond well to the *in vitro* environment, these procedures might sometimes involve *in vitro* seed germination. In other situations, different tissues may be employed in tissue culture to yield a large number of plants relatively quickly. To increase the number of rare or highly valued plants to share with other botanical gardens, plant propagators must have a thorough understanding of plant propagation. During propagation, a growing environment must be provided while taking into account the diminished ability to absorb water in the absence of a root system. Desiccation must be avoided until the plant is firmly established and able to withstand the problems of the real-world environment (Oliver et al., 2020). People have propagated plants for ages with little to no special equipment by completely understanding seasonal growth patterns, weather cycles, etc.

The essential facilities depend on the type of propagation. Because businesses may expand and frequently need to adapt to changes in consumer demand, facilities should be constructed to allow for maximum flexibility in terms of usage and expansion. Propagation is the most fundamental aspect of a greenhouse or nursery business; hence these areas must be close to the primary growing regions. The growth facilities can be used as extra space if needed for propagation. Propagation zones should be close to one another rather than dispersed around a facility for effective management. Nowadays, many nurseries and greenhouse operations have regional production facilities dispersed throughout several regions, allowing the business to benefit from certain climatic conditions or get better access to established or emerging markets. Most of the time, having a separate propagation space for each facility is preferable. It is simple to control propagation timetables and have transplants ready to shift into the production phase if they are produced locally because climates and local weather conditions vary, even within a given zone. Both the physical topography and the production requirements must be taken into account (Schimel et al., 1996).

2.3 PROJECT SETUP

Propagation facilities require a lot of water, so it is necessary to take some precautions to ensure that the site has appropriate water drainage away from it. Drainage problems are simple to fix before the building of structures, but they are challenging and expensive to fix afterward. Excessive surface water not only makes for uncomfortable working conditions, but it also acts as a breeding ground for pests like snails, insects, fungi, and pathogens that cause root rot like Phytophthora (Figure 2.2), Rhizoctonia (Figure 2.3), and Pythium species (Figure 2.4).

To ensure appropriate water removal, tiles should be put in areas with slopes of less than 1% (Horler et al., 1980). Some governments now control the amount of water that is discharged from greenhouses and nurseries. Stormwater runoff is managed via water catchments, which also lessen the quantity of nonpoint source pollution that leaves the site. The water in these catchment ponds is typically used by nurseries and greenhouses to irrigate their developing crops, not for multiplication.

Figure 2.2. Phytophthora. (Source:https://upload.wikimedia.org/wikipedia/commons/a/aa/Late_blight_on_potato_leaf_2.jpg, Creative Commons License)

Figure 2.3. Rhizoctonia. (Source:https://upload.wikimedia.org/wikipedia/commons/9/91/Rhizoctonia_solani.jpg, Creative Commons License)

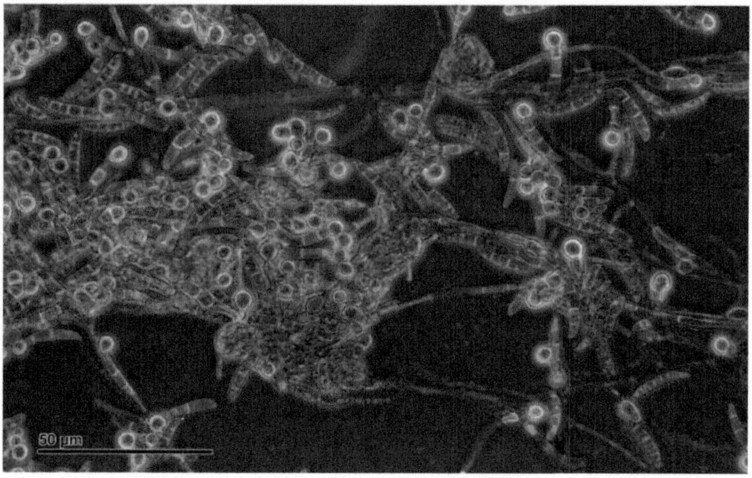

Figure 2.4. Pythium. (Source:https://upload.wikimedia.org/wikipedia/commons/f/f8/Pythium_%28257_23%29.jpg, Creative Commons License)

Assuring that the location has access to significant amounts of high-quality water is perhaps the most crucial factor to take into account before placing any horticultural operation. The same applies to propagation. Because young, sensitive plants are more vulnerable to stresses like high salts and pathogen attacks, water used for propagation must be of higher quality than that used to develop the crop. A site for a propagation facility should be chosen after testing the water quality. Salinity, alkalinity, pH, hardness, electrical conductivity, and mineral content are the key aspects of water quality.

If the quality criteria are the same, well water is preferred to surface water. During a growing season, surface water supplies are more unpredictable than well water because they are impacted by heavy rains, droughts, point, and nonpoint source pollution, problems brought on by algae, and disease-causing organisms. If surface water is the sole supply available, a multistage treatment plant might be required to assure high-quality water for propagation. To do this, water is run through a sand filter to get rid of as many particles as possible. The propagation house may also need in-line canister filters to keep the mist and fog nozzles from clogging. Either a chlorination system or UV lights can be used to disinfect surface water intended for use in propagation. Chlorination is currently the most popular and efficient method. Many municipalities use chlorine gas as a disinfectant, but because it is dangerous and difficult to handle, most propagators instead use sodium hypochlorite (NaOCl), which is an ingredient in household bleach products, or calcium hypochlorite ($Ca(OCl)_2$), which is the substance used to treat swimming pools. The quantity of organic residue, temperature, pH, and treatment duration are all factors that affect how much chlorine is needed to treat the water. The majority of water sources have a free residual level of chlorine at the output tap will be about 1 ppm, which is attained by adding 10 ppm at the treatment facility (Lonigro et al., 2017).

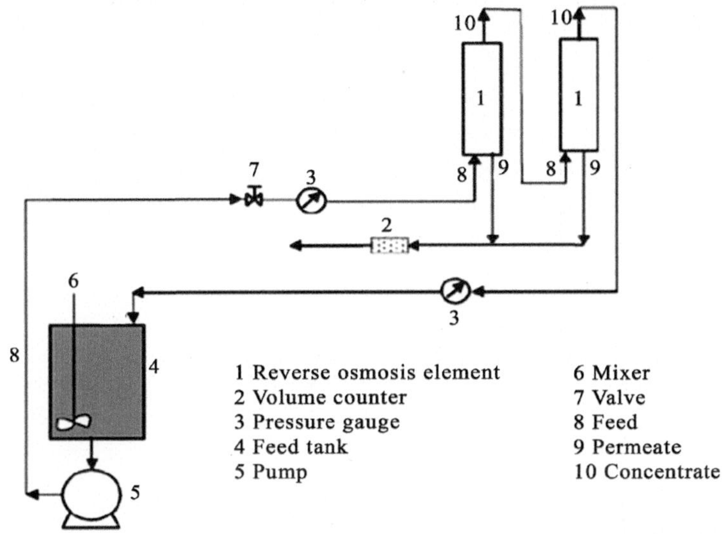

Figure 2.5. Reverse osmosis. (Source: Simonič M. Reverse Osmosis Treatment of Wastewater for Reuse as Process Water—A Case Study. *Membranes*. 2021; 11(12):976. https://doi.org/10.3390/membranes11120976, Creative Commons License)

A high alkalinity level is one issue that can be quickly fixed by adding acid to the water line. Reverse osmosis (Figure 2.5) treatment facilities can reduce high total dissolved solids and high electrical conductivity, two issues that are frequent in coastal or dry locations. Before choosing a place for propagation, it is crucial to investigate the economics of water treatment due to the significance of a sustainable, high-quality water supply. If considerable water treatment is required, municipal water supplies may be cost-competitive and can be used for propagation. Most communities add chlorine to the water supply to purify it, and it is completely safe to use for reproduction. To assist prevent tooth decay, the majority of towns now fluoridate their water supplies. Most plants are unaffected by the added fluoride, however several species of the Agavaceae (Dracaena and

Cordyline) and Liliaceae (Chlorophytum and Lilium) store fluorides in their leaf tissues, which can lead to necrotic leaf tips. Typically, this issue only arises when the fluoride level is higher than 1 ppm (Jacobson et al., 1966).

The type of propagation to be done determines the size and design of the propagation facility. The majority of the cultural needs will nearly always be met by greenhouses, but support facilities are also required. Trimming the cuttings when they are taken from the stock plants is one of the duties that can be performed in the field. The number of cuttings attached (planted) per area (plant density) and the period that the cuttings will occupy the propagation space all determine the amount of space needed for propagation. To make up for losses brought on by the low germination rates, and a variety of other biotic and environmental variables, more propagules may be planted. Good propagators pay close attention to the details and are constantly seeking ways to reduce loss during propagation.

Figure 2.6. Polyethylene-covered greenhouses. (Source: Dehbi, Abdelkader & Youssef, Boulos & Chappey, Corinne & Mourad, Abdel-Hamid & Picuno, Pietro & Statuto, Dina. (2017). Multi-Layers Polyethylene Film For Crop Protection In Harsh Climatic Conditions. Advances in Materials Science and Engineering. 2017. 10.1155/2017/4205862., Creative Commons License)

We must choose the density of the propagules in the propagation area and whether they will be in open beds or plots to calculate the amount of space needed for a specific crop. The most challenging species and cultivars are frequently grown in open bed systems because they offer the propagules the best drainage and aeration. Plant propagation is frequently done in greenhouses, which makes it simple to control all of the environmental factors that affect tissue propagation. Small growers might dedicate the majority of a greenhouse to production and only use a portion of it for propagation. Large growers have a variety of propagation-only greenhouses available to them. According to an early report, about 63 percent of the greenhouse area in the US is covered with polyethylene sheeting, 14 percent with glass, and the remaining 20 percent or so with stiff plastic panels and fiberglass-reinforced plastic (Boosalis & Scharen, 1959).

The most popular propagation infrastructure are polyethylene-covered greenhouses (Figure 2.6) because they are less expensive, double polyethylene provides better insulation, and even while light levels are less than under glass, they are still more than sufficient for propagation. Initially, woods like redwood or cypress that were resistant to rot were used to construct greenhouse frames. These later gave way to steel pipe frames holding the glass in place with wooden girdles. This technology was still in use after World War II when new building methods and materials started to become available. As opposed to the time-consuming job of removing wooden sills, and repainting every few years, aluminum components started to replace wooden frames. Galvanized steel that has been produced in a range of shapes and sizes is most frequently used to make greenhouse frames.

2.4 INFRASTRUCTURE

Cold frames and hot beds are the main structures used to propagate plants. These frames were constructed into the ground, with an elevation on the north side such that the glass girdle that covered the building had a 15° slope to the south to capture more sunshine in the winter (Hosikian et al., 2010). Although they were effective, they were challenging to automate. The limited air gap caused cold frames and hot beds to heat up or cool down quickly as the weather changed, requiring regular maintenance. They are used sparingly today for commercial reasons due to these restrictions and the amount of manual labor needed to examine and maintain the frames regularly. Gardeners continue to utilize cold frames and hot beds because they are simple to erect and affordable. The thermal properties of the covering and the amount of exposed surface area determine the cost of heating a greenhouse in a specific region. The height of the sidewall has increased as mechanization has grown in importance in greenhouses to provide room for the curtains, trolley, plumbing, lighting, and other equipment positioned above the growing area. The majority of farmers use significantly taller structures, some of which have sides as high as four meters. Some companies used different flat-roofed designs, while others maintained the classic contour of the old structures. To attain the open-roof goal, several manufacturers have used many many techniques (Soni et al., 2017). These houses often cost roughly the same as conventional buildings, depending on the design chosen.

Although retractable-roof greenhouses have not been extensively employed in propagation, this is likely to change in the future. They offer a more natural environment for plant growth, which benefits both propagators and growers. The plants are more hardened than the normal plants produced in a contained environment because they are exposed to the elements in a regulated manner, particularly wind, temperatures, and intense sunlight. Operational costs are comparable to those of other structures, particularly if the greenhouse is utilized with minimal heat. With this design, summer cooling expenses are eliminated. Maintenance expenses have been greater than with conventional greenhouses because of the increased use of electric motors, pulleys, cables, and other miscellaneous machinery essential to open and close the roof. As one moves closer to the equator, cooling greenhouse space becomes more crucial; conversely, heating becomes more crucial in northern latitudes. Temperatures within greenhouses are frequently 15° to 20°C warmer than the air temperature outside (Adams et al., 2001). This is not an issue in the winter, but in the late spring and summer, when warm conditions are present inside, most crops cannot grow because of the high interior temperatures. Some producers in warmer climes frequently cease production in greenhouses during the summer and, whenever practical, shift it to outdoor gravel beds. Natural ventilation was the conventional method for greenhouse cooling in colder regions. However, modifications to greenhouse architecture, particularly the development of poly houses, resulted in the replacement of natural ventilation systems with fan systems that draw fresh air into the greenhouse (Zobayed et al., 2000).

Passive cooling systems with fewer fans are more realistic now that the cost of electricity has increased. The generation of a strong, healthy plant that can be sold is the ultimate result of effective propagation, thus the technique that will be used as well as the environmental needs of the species being propagated are the main issues that are to be addressed. The availability of water and temperature are the main factors limiting plant dispersion on a global scale. These two factors must be taken into account the most when designing a facility for plant propagation. Most species like a warm root substrate, high humidity levels, and high plant-available water levels during the early stage of propagation. In addition to being a necessity for plant growth, light may also be necessary for some plant seeds to germinate. Airflow and oxygen availability are additional criteria that are crucial for the design of plant propagation facilities. Water that is available to plants can be supplied through a variety of irrigation techniques, including drip irrigation, automated misting systems, and manual watering. Mist bench systems use nozzles that produce a relatively high volume of water. Typically, one nozzle is needed for every 1.4 to 4.6 m^2 of propagation bench space, depending on the model (Eibl & Eibl, 2008). The mist can be delivered sporadically by simply installing a timer. Due to the low pressure needed, mist systems offer the advantage of having relatively affordable pumps; yet, the huge volume of water might result in higher disease pressure. For workers, the damp climate might be uncomfortable.

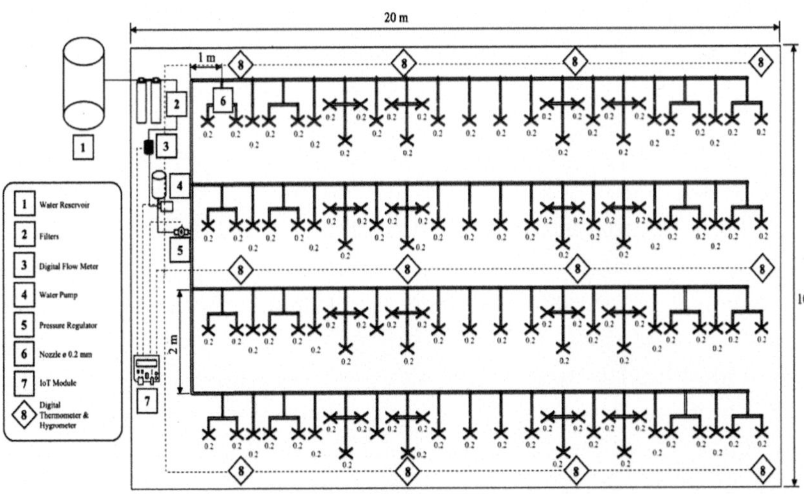

(a)

Figure 2.7. Fog systems. (Source: Eduard R, Ruslan W, Iskandar I, Setyanto D. Setting Temperature and Humidity with a Misting System in a Pilot Greenhouse at Cisauk-Tangerang, Indonesia. *Applied Sciences*. 2022; 12(18):9192. https://doi.org/10.3390/app12189192, Creative Commons License)

As an alternative, fog systems (Figure 2.7) can be utilized to maintain a cool temperature and increase humidity in the propagation area to lower plant transpiration losses. The water droplets evaporate very quickly. Much less water is needed for fog systems, but higher pressure and superior water quality are necessary and can increase long-term operational expenses. Fog systems, on the other hand, offer a very constant temperature throughout the building and, because of the quick evaporation rate, a relaxing working atmosphere for staff. The quick development of a new plant generation depends on temperature. Generally speaking, at a temperature between 21°C and 27°C, the majority of species can be successfully grown from seeds or cuttings. To warm the substrate and encourage root growth, heat is generated from below. Although hanging forced air unit heaters are another option, they are less preferable because they generate heat from above. A heating mat or heating coils can be used to supply heat for building requirements. These electric appliances are available in a variety of sizes and typically contain a thermostat to regulate the temperature. PVC frames can be used to support plastic barriers to lower light intensity or to increase humidity in the propagation area (Reddy & De Busk, 1985).

Cuttings of herbaceous, softwood, and semi-hardwood species can be vegetatively propagated in environments with intermittent mist systems. By increasing the relative humidity in the area around the cuttings and lowering leaf temperature through evaporative cooling, mist reduces water loss from the cuttings. It is possible to build an intermittent mist system inside a greenhouse to regulate the temperature, day length, and other environmental factors. Intermittent mist systems, however, can also be built outside. If the plant material being propagated is acclimated to greater light intensities and temperatures, this is frequently advantageous. Intermittent mist systems need a source of high-quality water under pressure of 40 PSI and energy, whether they are built inside or outside. The day/night clock in the fundamental intermittent mist system switches the system on during the day and off at night. A second clock regulates how frequently and how long misting events last when the system is turned on during the day. Typically, this second clock is configured so that misting lasts for 5 to 10 seconds once every 5 to 10 minutes. However, studies have shown that when misting is kept to a minimum, many plants grow adventitious roots more quickly. Since time clocks are insensitive to changes in their surroundings, it is challenging to accomplish this with them. For instance, mist systems regulated by a time clock continue to operate as if it were a hot day on cloudy or rainy days when the light intensity and temperature are lower. Time clocks have thus been replaced with different mist-controlling technologies that adapt to changes in the environment as alternatives.

Figure 2.8. Ornamental plants.

A breeding programme of *Hippeastrum* × *chmielii* at Warsaw University of Life Sciences, Poland. (a) Seed germination on petri dishes; (b) hybrid seedlings in their juvenile phase under greenhouse conditions; (c) flowering of the seedlings' population en masse in the third year after sowing. (Source: Marasek-Ciolakowska A, Sochacki D, Marciniak P. Breeding Aspects of Selected Ornamental Bulbous Crops. *Agronomy*. 2021; 11(9):1709. https://doi.org/10.3390/agronomy11091709, Creative Commons License)

The majority of nurseries depend heavily on the propagation of nursery and ornamental plants (Figure 2.8). Commercial propagation techniques have undergone significant change over the past 25 years (Murashige, 1974). One of these developments is the decreased usage of soil nowadays. Using soil-less media in high-density systems, such as plug trays or cell packs, nursery, and ornamental plants are increasingly being grown from seeds and explants. Advantages of using containers include flexibility in planting times, simplicity in handling and moving materials, and adaption.

Modern computer-controlled greenhouses (Figure 2.9), playhouses, or other secure environments that offer the necessary moisture, temperature, and light are often where germination and rooting occur. These innovations make it possible to propagate plants in methods that are more effective, affordable, and labor-saving. Although rooting media typically have a coarser texture than media for sowing seeds, both types can be compared based on their practical qualities. Suitable air-moisture balance, support, and anchoring are all provided by a spreading medium for metabolic processes.

For plants to grow that are acceptable for transplanting or sale, the medium must be stable enough to permit and maintain adequate anchoring and support.

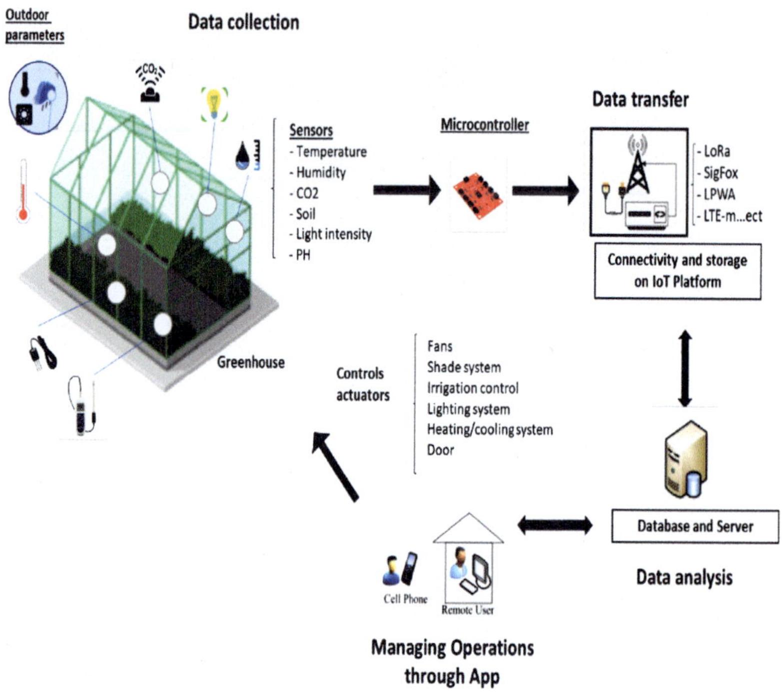

Figure 2.9. Computer-controlled greenhouses. (Source: Bersani, C., Ruggiero, C., Sacile, R., Soussi, A., & Zero, E. (2022). Internet of Things Approaches for Monitoring and Control of Smart Greenhouses in Industry 4.0. *Energies*, *15*(10), 3834. https://doi.org/10.3390/en15103834, Creative Commons License)

2.5 MEDIUM

The optimal propagation medium needs to be loose or friable enough to preserve sufficient air for germination and cell division at the base of the cutting and the development of new roots. Starches, lipids, and other food reserves in the seed or explant must be oxidized by respiration, and the amount of oxygen used depends on the level of metabolic activity. Adequate aeration of the medium enables the elimination of carbon dioxide, a by-product of respiration that, when present in excess, can harm seedlings. It is important to keep the medium moist during propagation while also maintaining a humid environment in the surroundings. The medium around a seed base needs to be moist at all times. To prevent seeds, cuttings, or new tissues from drowning in water, it should have a high moisture-holding capacity while allowing extra water to drain freely. If the medium is not adequately drained, roots may rot. The success of propagation can be greatly influenced by a variety of factors, including species, the season, the application of growth-promoting agents, or the interaction of these and/or other variables. The starting media must be free of diseases, pests, weeds, and toxic chemicals, whether it is for seeds, cuttings, or transplants. Early problems, such as the presence of weed seeds, will result in more labor costs and expenses for removal. The medium should typically have few or no nutrients. Before the formation of roots, nutrients are of little use and can be supplemented with water as required. Nutrients can harm seedlings or cuttings if they are present in excess.

2.6 PARAMETERS

Physical parameters like bulk density and pore sizes and chemical properties like electrical conductivity and pH can be used to technically describe medium characteristics. The propagator can analyze and compare several batches of the same mixture using these attributes as a basis, which helps them gain a better understanding of media dynamics. Electrical conductivity (EC), which gives an indirect measurement of the nutritional or fertility status of a material or medium is used to quantify soluble salts (Dorai et al., 2001). Inorganic ions or nutrients that have been dissolved in water are known as soluble salts. If managed properly, certain waste derivatives, such as municipal composts and paper mill sludges with mediate salt levels, can be used as rooting media. Salts leach quickly in environments containing these minerals. There is a significant risk of burning the cuttings because the salts could leach insufficiently or inconsistently. Before use, always check the soluble salt content of the medium made from waste. We need to try any new material in a small setting first to see how it performs before applying it widely. Use no more than 30% of the material by volume in the mixture as a general rule. By itself, using this lesser percentage frequently lessens the potential damage brought on by salts or other hazardous compounds. Cuttings or seedlings are extremely vulnerable to salts in the media. Cuttings from woody species that root quickly appear to be less sensitive to high salt levels than those that root slowly, whereas cuttings from herbaceous species appear to be more sensitive. While there is some data on the relative resiliency of ornamental plants to salt concentrations, there is scant or no conclusive data on the resiliency of seedlings or cuttings during propagation.

Figure 2.10. Rhododendrons. (Source: Kang L, Zheng K, Xie Y, Deng Y, Yu Y, Zhu M, Xi R, Deng X. Efficient Tissue Culture Protocol for *Magnolia lucida* (Magnoliaceae) and Confirmation of Genetic Stability of the Regenerated Plants. *Plants*. 2020; 9(8):997. https://doi.org/10.3390/plants9080997, Creative Commons License)

Figure 2.11. Azaleas. Japanese azalea: (A,B) Photograph of "Kinshū-makura" (A) and 'Misome-guruma' (B), edited in 1692. (C–E) Flower photos

of *Rhododendron kaempferi* 'Nikkō-misome' after just flowering (C), approximately 50 days (D), and approximately 100 days (E). (Source: Cheon K-S, Nakatsuka A, Tasaki K and Kobayashi N (2018) Long-lasting Corolla Cultivars in Japanese Azaleas: A Mutant AP3/DEF Homolog Identified in Traditional Azalea Cultivars from More Than 300 Years Ago. Front. Plant Sci. 8:2239. doi: 10.3389/fpls.2017.02239, Creative Commons License)

Rhododendrons (Figure 2.10) and azaleas (Figure 2.11) (*Rhododendron* spp.) and other Ericaceous species are acidophilic plants that require an acidic medium with a pH range of 3.8 to 4.5 to support growth, however rooting can occur at higher pH levels (Zimmerman, 1985). Numerous non-ericaceous, deciduous, and evergreen plants have been established in conditions with pH ranges of 4.0 to 8.4 (Huetteman & Preece, 1993). The pH range of 5.0 to 8.4 has not been associated with any negative consequences. Common shrubs are amphi-tolerant or tolerant of a wide range of pH levels, therefore pH change may not be a significant barrier to germination as was previously believed. The solubility or availability of nutrients is often the main factor to take into account when determining pH. The goal of mixing is to generate or produce media with specified ranges of physical and chemical properties. However, organic materials with biological qualities, such peat, bark, composts, and by-products like paper mill sludges, may be helpful to seeds or explants during multiplication. Materials of this type typically consist of organic matter or plant residues in various stages of decomposition, but they may also contain beneficial microorganisms that produce antibacterial and antifungal agents, root-promoting substances like auxins and cytokinins, and/or biocatalysts like enzymes, vitamins, and antibiotics. It is essential to know the presence of these compounds in wastes and composts as well as their ability to promote roots. The use of compost in propagation specifically to take use of these biological qualities seems to be less studied.

The majority of propagation media are combinations of different substances. While media can occasionally be produced using three or even four different materials, most of the time, media with equivalent qualities can be produced with just two. These are simpler to prepare, and their performance or result may be more predictable. Propagation or culture media have been made from a wide variety of materials. These can be classified as organic or inorganic and used largely in their natural state, through production, or blending. The following are examples of inorganic materials: sand, perlite, vermiculite, polystyrene, and rock wool or mineral wool. Peat, bark, wood scraps, and composts are examples of organic materials, while many organic or farm-derived waste materials have only been employed sparingly. Typically, the mineral component is utilized to increase the proportions of air-filled pore space to improve drainage and aeration. In the last 25 years, the use of sand has drastically decreased, and lighter, easier-to-handle materials like perlite or commercial premixes have mostly taken their place. Sand is heavy, and because it needs to be disinfected to control viruses and pests, transport expenses can be significant. Because finer sand lowers pore space and drainage, a medium grade of washed sand is advised. Sand is frequently employed in combinations with peat, perlite, and/or other organic components, and occasionally as a single-component media, however, it contains little to no nutrients. In sand and peat combinations with peat proportions over 50%, several varieties of explants will decay due to conditions brought on by excessive moisture (Maene & Debergh, 1985).

Figure 2.12. Perlite in tissue culture. (Source: Debnath, S. C., & Goyali, J. C. (2020). In Vitro Propagation and Variation of Antioxidant Properties in Micropropagated Vaccinium Berry Plants—A Review. *Molecules*, *25*(4), 788. https://doi.org/10.3390/molecules25040788, Creative Commons License)

2.7 INSULATION

Perlite, (Figure 2.12) which was first employed as insulation, is one of the most commonly utilized materials in culture mediums. It is sometimes used alone but is more frequently combined with sphagnum peat in mixes of 2:1,

1:1, and 1:2 by volume (Bedini et al., 2012). Aluminum silicate volcanic rock that expands and extrudes when heated quickly to 1000°C is mined to produce perlite. It is non-water or nutrient-holding, white, and light, with irregularly shaped particles and sealed interior air spaces. Its principal applications involve preserving or expanding pore spaces and replacing sand with something lighter. It may be used again if reinfested, is sterile, does not decompose, and drains readily. The pH of perlite ranges from 6.5 to 8.4 but is often near neutral. Typically, a coarse grade is advised for use in propagation.

Figure 2.13. Vermiculite. (Source:https://upload.wikimedia.org/wikipedia/commons/b/b0/Vermiculite1.jpg, Creative Commons License)

Vermiculite, (Figure 2.13) another heat-treated mineral, is extracted from an ore that resembles mica and contains aluminum, iron, and magnesium. It has a pH that typically ranges between 7.0 and 7.5, but can range from 6.0 to 9.5 depending on the source. It is sterile, light, and similar to perlite (Ghosh et al., 2018). It behaves like an organic substance and is very good at retaining nutrients and water. It has a comparatively high Ca, Mg, and K content. When moistened, it is easily compressed and does not expand. This technique is more frequently used with explants that root quickly, such as chrysanthemum and coleus (*Coleus* sp.). Fine vermiculite and peat moss (1:1 by volume) were employed to enhance the roots of sugar maple (Couselo et al., 2012). Vermiculite is frequently included in germination mixtures along with peat, bark, and composts because it is thought to be an effective ingredient (Figure 2.14).

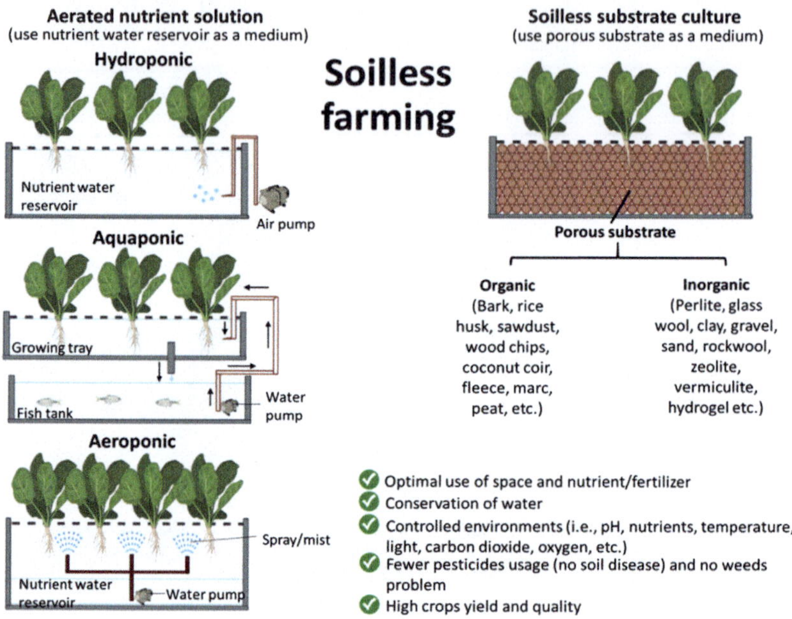

Figure 2.14. Overview of alternate materials. (Source: Maluin FN, Hussein MZ, Nik Ibrahim NNL, Wayayok A, Hashim N. Some Emerging Opportunities of Nanotechnology Development for Soilless and Microgreen Farming. *Agronomy*. 2021; 11(6):1213. https://doi.org/10.3390/agronomy11061213, Creative Commons License)

As a less expensive alternative to perlite, polystyrene flakes and beads have been employed. It is inert, light, and close to pH neutral like perlite. Since particles tend to float or migrate to the top of the medium, it is challenging to mix well. The overall pore space volume of rock wool, another inert, sterile, heat-treated substance, is extremely high at 98 percent, and it has a high aeration porosity of roughly 50 percent. It drains freely and cannot hold nutrients or water in its structure. This makes it the perfect substrate for establishing explants from succulent plants like coleus (Figure 2.15), begonia (*Begonia* spp.) (Figure 2.16), and impatiens (*Impatiens* spp.) (Figure 2.17).

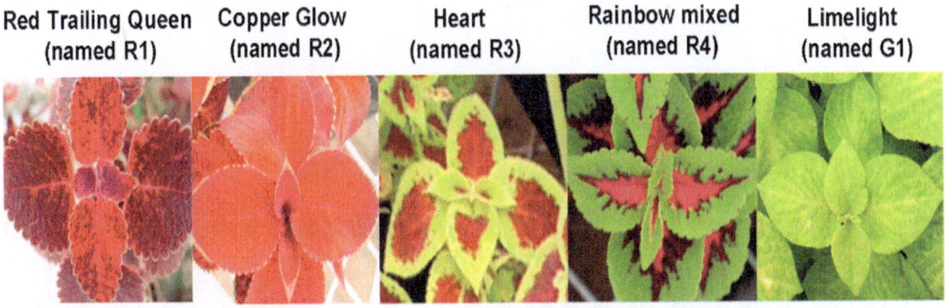

Figure 2.15. Coleus. (Source: Zhu Q, Xie X, Lin H, Sui S, Shen R, Yang Z, Lu K, Li M, Liu Y-G. Isolation and Functional Characterization of a Phenylalanine Ammonia-Lyase Gene (*SsPAL1*) from *Coleus* (*Solenostemon scutellarioides* (L.) Codd). *Molecules*. 2015; 20(9):16833-16851. https://doi.org/10.3390/molecules200916833, Creative Commons License)

Figure 2.16. Begonia. (Source: https://commons.wikimedia.org/wiki/File:Begonia_grandis1.jpg, Creative Commons License)

Figure 2.17. Impatiens. Flower specimens of the *Impatiens walleriana* studied in an article (a): orange petals and (b): pink petals (Source: https://commons.wikimedia.org/wiki/File:Starr_070906-8887_Impatiens_hawkeri.jpg, Creative Commons License)

Figure 2.18. Peat. (Source: Topcuoğlu, B. , & Turan, M. , (Eds.). (2018). Peat. IntechOpen. https://doi.org/10.5772/intechopen.69565, Creative Commons License)

To be utilized as a single-component medium, it is often made into a variety of prefabricated shapes, including cylinders, blocks, and wedges. The qualities of rock wool are dependable and consistent, and it is strong. These plastics have great water-holding capabilities and an almost neutral pH. Foams or synthetic materials, like rock wool, are typically produced and supplied as premade blocks, cubes, or wedges in ready-to-use tray plugs. They need to be handled cautiously because they are fragile and easily destroyed. They cannot regain their former shapes after being crushed or compacted.

Peat (Figure 2.18) is a partially degraded by-product of mosses that are typically found in swamp environments. It is widely distributed in regions with cold climates like Canada, Northern Europe, and Russia. Peat comes in a wide variety of types, grades, and qualities. The type of peat frequently used as media is sphagnum peat. Over 75% of it is *Sphagnum* sp. and at least 90% of it is organic matter. Young sphagnum peat has a pH between 3.5 and 4.0, which is extremely acidic (Álvarez et al., 2017). It is reasonably devoid of infesting organisms, light like perlite, has an aeration porosity of about 30%, and contains little or no nutrients. Older, darker peats typically have higher pH levels, are denser, and are more decomposed. Sphagnum peat decomposes slowly, can have fungistatic properties similar to its parent material, sphagnum moss, and can absorb between 10 and 20 times its weight in water (Rice & Currah, 2006). However, these qualities are frequently lost during decomposition.

Figure 2.19. Pine bark. (Source: https://commons.wikimedia.org/wiki/File:Bark_of_an_adult_pine_%28Pinus%29_tree.jpg, Creative Commons License)

Pine bark, (Figure 2.19) a by-product of the industry, and composted pine bark are frequently used as media by some nurseries and in container growing media. It is believed to have disease-suppressing effects, is typically low in salts, and is quite light. Pine bark pH ranges from 3.8 to 6.5, depending on the source, while those with pH readings between 5.0 and 5.5 are more frequently employed. It is a good alternative to peat, but it holds less water. Water can freely drain through the bark due to its open structure. Although their use has been increasingly restricted, related by-products like sawdust and wood chips have been employed in the industry. Many of these items break down while being used, which occasionally causes the nitrogen level to drop. Media mixed with sawdust, in particular, may get waterlogged because of its exceptionally high capacity to retain water.

Figure 2.20. Floriculture. (Source:https://upload.wikimedia.org/wikipedia/commons/thumb/3/37/Floriculture.jpg/1200px-Floriculture.jpg, Creative Commons License)

2.8 FLORICULTURE

Over the past ten years, the floriculture (Figure 2.20) business has seen a significant increase in the usage of coir, a waste product left over from the production of coir fiber from coconut husks. Some growers state that it functions better as a growing medium than peat. Although it has an open structure, it is slow to degrade and compaction resistant. It has a relatively high water-holding capacity, but compared to peat, it is less likely to contract diseases. The pH and soluble salt concentration vary greatly depending on the source. The use of coir in mixtures with perlite between 15 and 75 percent by volume has produced excellent rooting results. Composts, leaf molds, coir, paper mill sludges, and other waste-derived materials are receiving more and more attention as media. Inconsistent quality, potential phytotoxicity caused by high salt levels or unfavorable pH, variations in species response, and a lack of scientific knowledge are major barriers to using these materials in propagation. In addition, compost or a by-product from a paper mill may not always be more cost-effective than more conventional materials like perlite or peat or be accessible within a reasonable transporting distance. Various paper mill sludges and composts have been examined, and it turns out that they are three to five times heavier and have porosity and aeration qualities that are on par with or better than peat. As a result, these waste-derived materials make excellent peat substitutes in media.

Many woody species have been successfully rooted using volume fractions of up to 60% and 75%, respectively, of paper mill sludges and municipal waste composts. When these components are combined with perlite rather than peat, the results are typically superior. Similar to peat, the sludges and composts have a great capacity to retain water, which tends to make the mixtures wet. Explants and seed propagation can both be significantly impacted by the media utilized for propagation. It's interesting to note that neither the germination of seeds nor the rooting of explants requires a medium. Explants have been commercially planted using aeroponics and will develop roots in humid settings. In this instance, a nutritional solution is sprayed as they are suspended in the air. Similarly, seeds can easily germinate in Petri dishes that have been lined with filter paper or paper towels. Cuttings can also be rooted in a glass of water; however, this is not advised because the low oxygen content leads to the formation of poor-quality roots. Although it improves roots, air bubbles in the water are not always feasible. The propagator must create the ideal environment for seeds or explants to successfully germinate. Sufficient amounts of water and oxygen, the right temperature, and a pH in the right range must be provided.

The processes of germination and roots typically do not require the environment to provide light or nutrients. Light may be required for some species to emerge from dormancy, although it has little to no effect on the germination of many other species. The seed or explant already has the necessary nutrients, therefore additional nutrients are not needed for germination, roots, or the development of accidental shoots if the stock plants were treated appropriately. However, nutritional fertilization is required for continued growth. Understanding propagation outcomes will significantly improve as a result. To provide a disease-free plant for the consumer, disease management during plant propagation is crucial. The fast spread of disease may be explained by the conditions in which seedlings are cultivated, which are essentially in monoculture. High relative humidity conditions, which favor many foliar diseases, are frequently used to grow vegetatively produced explants. Cold-season plants are more susceptible to some root rot diseases in many parts of the world due to exposure to high temperatures. Anyone involved in the production of plants must understand that growing a strong, marketable plant begins before and continues throughout plant propagation. We need to establish the groundwork for crop health during this stage of plant development. Although seedlings or rooted explants may not escape from contagious diseases, they may become malformed, discolored, stunted, or less resilient in general. Numerous plant diseases that develop during plant reproduction have the power to lower the grade or quality of affected plants. Some infections that affect plants that have been propagated vegetatively have not been found in greenhouses, nurseries, or garden centers. This frequently happens when the symptoms of diseases, such as nutrient shortages, pesticide phytotoxicity, or harm from abiotic pressures, are misdiagnosed or mixed up with those of other diseases.

2.9 PATHOGENS

Plant pathogens, such as fungi, fungi-like organisms, bacteria, mollicutes, viruses, viroids, and nematodes, can infect plants in propagation systems and cause disease. Fungi are eukaryotic and lack chlorophyll. They typically have cell walls formed of chitin and other polymers and are filamentous, branched organisms that reproduce through spores (Leifert et al., 1991; Leifert & Waites, 1992). The fungi-like organisms, including Phytophthora and Pythium, are part of a group of microbes that can cause downy mildew, damping-off, root rots, wilting rots, and other extremely harmful diseases. Oomycetes are a class of pathogens that resemble fungi greatly in that they develop in filaments and reproduce by spores. These species, which were previously thought to be fungi, are now believed to be more closely related to the brown algae. Soft rots and leaf spot diseases are frequently brought on by prokaryotic organisms, like bacteria, which are frequently present in propagation systems. Bacteria are single-celled organisms without a nucleus or organelles bound by two membranes, although they do have a hard cell wall. Bacteria that cause plant diseases typically have a rod or filamentous form. Mollicutes are smaller than bacteria and lack a cell wall, which separates them from bacteria. The plasma membrane encloses them. These species can take on a wide range of shapes because they lack a solid cell wall. Mollicutes are the cause of plant diseases that cause yellowing. Since they are much smaller than bacteria, viruses, and viroids cannot be detected with a light microscope. Both cannot replicate independently since they both depend on the host replication apparatus for multiplication. Viruses are nucleoproteins made of protein-coated nucleic acid (DNA or RNA). Viroids are essentially naked strands of RNA without a protein coat; however, they share many characteristics with viruses.

Figure 2.21. Tomato spotted wilt virus. (Source:https://upload.wikimedia.org/wikipedia/commons/7/72/Tomato_with_Tomato_Spotted_Wilt_Virus.jpg, Creative Commons License)

One of the main risks to good ornamental crops is plant viruses. Viral diseases can cause a wide range of symptoms, such as ringspots, mosaics, stunting, stem cankers, leaf spots, etc. Viruses can spread through a variety of channels.

Tomato spotted wilt virus (Figure 2.21) or TSWV and impatiens necrotic spot virus (INSV) (Figure 2.22) are frequently propagated when vegetative explants are obtained from infected stock plants. Tobacco mosaic virus (TMV) may be mechanically spread by handling plants with tools or hands contaminated with the virus. Insects can potentially act as a vector for viruses. Cauliflower mosaic virus or CMV is frequently spread by aphids, whereas INSV and TSWV are frequently spread by western flower thrips.

Figure 2.22. Impatiens necrotic spot virus. (Source:https://upload.wikimedia.org/wikipedia/commons/c/cc/INSV.JPG, Creative Commons License)

To guarantee the health of stock plants, it is crucial to get vegetative explants from a horticultural provider that has a viral indexing program. WFT can quickly obtain INSV and transmit it to healthy plants. A greenhouse's interior and outside need to be managed for weeds. For viruses and insects that serve as their vectors, weeds may behave as asymptomatic hosts. Plant seedlings from vegetatively produced crops separately, if possible, in several greenhouses. A virus might always enter the greenhouse through cuttings made vegetatively and then spread to seedlings by insects like WFT. Around the world, roses are frequently affected by rose mosaics. Apple mosaic virus (AMV) (Figure 2.23) and Prunus necrotic ringspot virus (PNRSV) (Figure 2.24) are the causes of rose mosaic (Figure 2.25). These viruses can exist simultaneously or separately and are frequently transferred when roses are propagated vegetatively. Vein clearing, mosaics, and line patterns, also known as oak leaf symptoms, are common symptoms in roses. Infected roses may show no symptoms in hot weather. Roses with rose mosaic infection may have smaller and fewer flowers.

Figure 2.23. Apple mosaic virus. (Source:https://upload.wikimedia.org/wikipedia/commons/thumb/c/c0/Pana%C5%A1ov%C3%A1n%C3%AD_listu_jablon%C4%9B.JPG/1200px-Pana%C5%A1ov%C3%A1n%C3%AD_listu_jablon%C4%9B.JPG, Creative Commons License)

In propagation systems, bacterial diseases like soft rot, bacterial leaf spot, bacterial wilt, and crown gall are frequently observed. In warm weather, soft rot from *Erwinia* spp. can convert explants beneath mist into a slimy, rotten mass (Joko et al., 2014). When propagating, the poinsettia is extremely prone to bacterial soft rot. Plants propagated by vegetative cuttings are susceptible to soft rot caused by *Erwinia carotovora* and *E. chrysanthemi*, respectively (Nieves–Brun, 1985). Wilting and collapse are typical signs. Tropical vascular rot and discoloration may also be linked to *E. chrysanthemi*.

Figure 2.24. Prunus necrotic ringspot virus (PNRSV)-infected *Cucumis sativus* (left) and healthy plant (right). (Source: Dong Z, Zhan B, Li S. Selection and Validation of Reference Genes for Gene Expression Studies Using Quantitative Real-Time PCR in Prunus Necrotic Ringspot Virus-Infected *Cucumis sativus*. *Viruses*. 2022; 14(6):1269. https://doi.org/10.3390/v14061269 , Creative Commons License)

Figure 2.25. Rose mosaic. (Source: By Malcolm Manners from Lakeland FL, USA - Rose Mosaic disease, CC BY 2.0, https://commons.wikimedia.org/w/index.php?curid=60523016 , Creative Commons License)

The infections that cause leaf spots on ornamental crops are caused by several microorganisms. Zinnia leaf spot caused by bacteria is one of the most prevalent. *Xanthomonas campestris pv. zinnae* is a bacterium that can be discovered on seeds and spreads to seedling leaves shortly after germination. Dark brown to black dots can be found on leaves. The bacterium spreads to nearby plants through overhead irrigation. By cleaning seeds before sowing, this specific disease can be avoided. Geranium bacterial blight, which is brought on by the bacterium *Xanthomonas campestris pv. pelargonii*, is one of the most harmful bacterial infections. It can also result in V-shaped lesions, wilting, and mortality in addition to leaf spots. It can spread quickly among zonal and ivy geraniums and is extremely contagious. It is best to get nodal cuttings from vendors who test their stock plants for this disease (Graham, 1964; M. K. Kim et al., 2007; Norman et al., 2006).

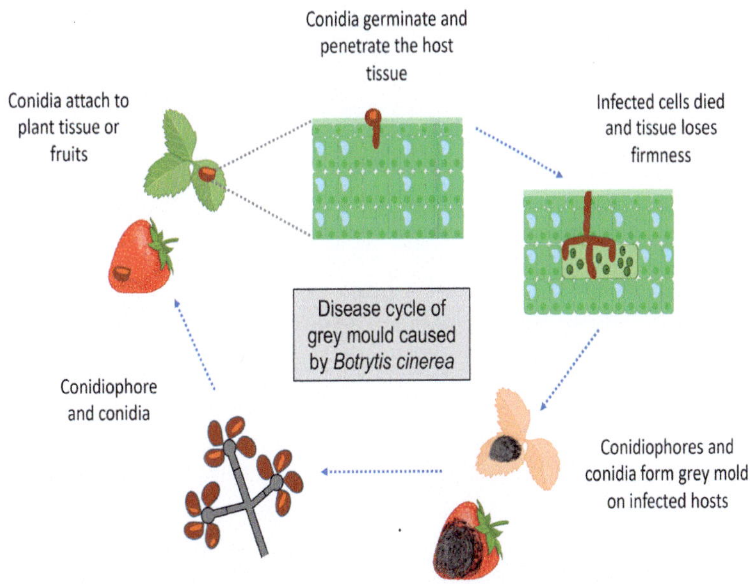

Figure 2.26. Botrytis blight.

(Source: Orozco-Mosqueda, Ma. del C., Kumar, A., Fadiji, A. E., Babalola, O. O., Puopolo, G., & Santoyo, G. (2023). Agroecological Management of the Grey Mould Fungus Botrytis cinerea by Plant Growth-Promoting Bacteria. *Plants*, *12*(3), 637. https://doi.org/10.3390/plants12030637, Creative Commons License)

The most common ornamental disease is botrytis blight, (Figure 2.26) often known as gray mold. *Botrytis cinerea* prefers moist environments, and leaf moisture is crucial for infection. The most common plant parts to sustain damage are older leaves or flowers, although any part of the plant is susceptible to damage in the optimum environment of stagnant air and excessive humidity. The blight caused by botrytis can be especially harmful to begonia, poinsettia, and geranium plants. In most greenhouses, *Botrytis cinerea* can be found on the tissue of dead plants as a saprophyte. It easily sporulates on plant tissue, and practically any activity in the greenhouse may cause spores to be released. For spores to begin to grow, free water must touch them. One of the more prevalent root rot diseases affecting floral crops is pythium root rot. It is present in many plants, including chrysanthemums and poinsettias. Wilting is among the early indicators of a problem. Plants with the infection may wilt quickly and die. Pythium is a water mold that can spread in irrigation water or be disseminated like Rhizoctonia. High quantities of soluble salt in the media encourage pythium root rot damage. As opposed to pythium root rot, Phytophthora crown or root rot is less frequent.

2.10 CONTAMINATION

Media that have been polluted by dirt or that have been incorrectly preserved can disseminate these fungi-like organisms. Stems and leaves with Phytophthora infection may be dark brown or black. A root rot that penetrates the stems of infected plants is one of the symptoms. Both phytophthora and pythium root rots are favored by overwatering, saturated, and poorly drained media. The sudden death has been linked to *Phytophthora ramorum* in coastal California is regarded to be an invasive pathogen in the United States. As more information about this disease is discovered, the host range is expanding. Rhododendron, Viburnum, mountain laurel, Pieris, and camellia are typical hosts. Pansies are susceptible to black root rot (*Thielaviopsis basicola*), which can also affect Fuchsia, Verbena, Vinca, petunia, and other plants. Plants that are infected typically exhibit poor color and limited development. Although they are uncommon in greenhouses, fungus leaf spots can develop on potted plants and rooted cuttings. Erysiphe, Microsphaera, and Podosphaera, three fungi that produce powdery mildew, are rarely found infecting plants in mist propagation systems because free water prevents spore development. While powdery mildews do not require free moisture to infect, high relative humidity is required for the pathogen to develop.

The organisms that cause stem rots, root rots, and foliar blights, Pythium and Phytophthora, are really more closely related to one another than the fungi that cause downy mildew. On a range of plants, including roses, impatiens, salvia, and snapdragon, downy mildew is growing more prevalent. Cool, rainy weather is favorable for infection. On herbaceous plants, the signs include leaf necrosis, foliar chlorosis, and stunting. Severe defoliation can happen to woody plants like roses. An angular brown, black, or maroon leaf spot may first emerge on the foliage in the early stages. Additionally, infected shoots may get a systemic infection and wilt quickly. On the lower side of the leaves, downy mildew is frequently noticeable. Rust diseases can affect perennial plants including Potentilla, as well as snapdragon, geranium, and snapdragon. The fungus is typically sporulating by the time yellow leaf spots are observed, releasing millions of yellow-to-orange or brown spores in pustules that make the fungus identification simple. Fungal growth may be seen on the basements of leaves, and symptoms include angular leaf patches and yellowed foliage. To eliminate germs and fungus on benches, pots, flats, tools, and sidewalks, we need to use disinfectants containing ammonium chlorides, such as Physan, Consan, and Green Shield.

To store plant waste, until it can be destroyed, each greenhouse needs a garbage can with a lid. Because soilless media aren't sterile, you might once in a while find traces of plant pathogens like Rhizoctonia, Thielaviopsis, and Pythium. Media should be stored and mixed on a clean table or concrete pad to prevent dirt contamination. Hose nozzles should never touch the ground to avoid contaminating clean medium with plant diseases. We need to take explants from healthy, disease-free plants. If possible, take fruit tree cuttings from a block of trees that has been confirmed free of common plant viruses. Purchase cuttings from horticulture vendors who test their stock plants for bacterial blight and INSV. The extra cost is justified by the savings from not having to deal with these diseases. We need to examine all newly arrived plants for any evidence of insects or disease. If we take the time to examine the roots of a few different plants, black root rot can be quickly identified. To lower humidity levels, vent moisture-laden air in the late afternoon and warm entering air. This is important because it can stop condensation from developing on leaf surfaces at night when Botrytis or downy mildew infections are most prone to happen. To lessen condensation on plastic greenhouse covers and leaf wetness, employ horizontal airflow fans. Install fans immediately above the crop to create a racetrack-like airflow. Avoid overhead irrigation if at all feasible, especially on crops that are vulnerable to botrytis blight. Botrytis and bacterial blight can be lessened by drip irrigation. Plants should be placed apart to allow for air circulation and light penetration through their crowns. Plants that are close together lose their lower leaves, which Botrytis quickly colonizes. High soluble salt concentrations encourage pythium root rot. Plant disease clinics have found that the absence of nutrient monitoring is responsible for at least 50% of the issues with floral crops they have identified. Ensure that stock plants receive the right amount of nutrients and are well-maintained. Although it is not required for seed germination or rooting cuttings, fertilizer will be required soon after. It is important to appoint at least one person to look out for disease, insects, and the earliest indications of poor plant growth. Early disease detection of affected plants is an effective management practice. Insects are a source of plant infections that harm floral crops.

CHAPTER 3

Primary and Secondary Metabolism

CONTENTS

3.1 INTRODUCTION

Research demonstrate that a variety of plant cells can be modified to develop photoautotrophically, and that one cotton line (COT-PA) can even develop in the presence of atmospheric CO2 (Widholm, 1989). However, the investigations have not yet provided an explanation for why only one cotton line can survive in low CO_2 environments and not the others. Since activity is only observed in the soybean lines SB-P and SB 1-P, neither of which would grow in ambient CO_2 conditions, carbonic anhydrase was ruled out as the reason. These soybean lines, which were started from various genotypes at various dates, have several traits in common. In terms of CO_2 fixation enzyme levels, the use of protease inhibitors has boosted the RuBPcase activity (Abdulbaki, Alsamadany, Alzahrani, & Olayinka, 2022). When compared to the cultures from C_3 species, the C_4 species (Amaranthus) (Figure 3.1), ACR-P and APO-P, do not exhibit enhanced PEPcase activity.

Figure 3.1. Amaranthus. (Source: https://upload.wikimedia.org/wikipedia/commons/thumb/4/43/3836_-_Amaranthus_caudatus_%28Zieramaranth%29.JPG/800px-3836_-_Amaranthus_caudatus_%28Zieramaranth%29.JPG, Creative Commons License).

When measured on a ChI basis, dark respiration is higher than mature leaves in cultures with low ChI levels, but it is not higher when calculated on a fresh weight basis. Research of the fluorescence emission spectra reveal that the ChI antenna systems are comparable to those of isolated chloroplasts and green algae, and in contrast to studies with leaves, no spectral distortion was seen. These cultures are currently a rich source of material for several genetic, molecular, biochemical, and physiological studies. As the primary trans locatable carbon source in the entire plant, CO_2 is digested to sucrose. Phloem and cell sap contain sucrose at quantities of about 10–25%, which is significantly greater than the typical concentrations in tissue culture medium(Brown & Shelp, 1997).

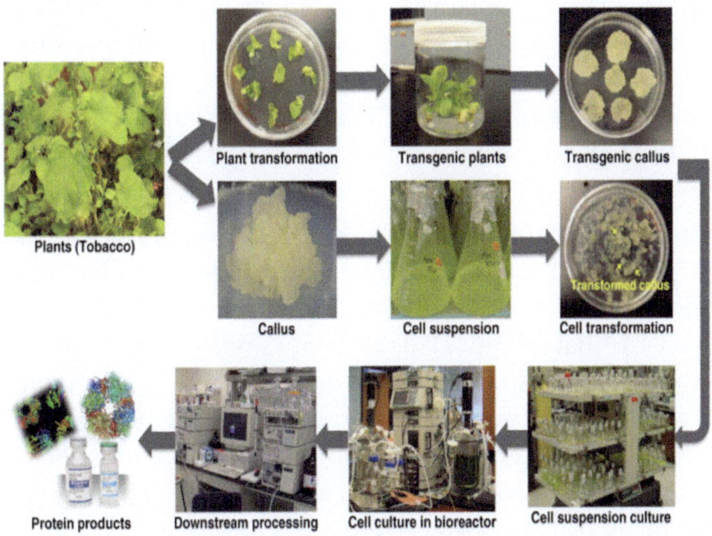

Figure 3.2. Plant cell suspensions. (Source: https://ars.els-cdn.com/content/image/1-s2.0-S0734975011000036-gr2.jpg, Creative Commons License).

3.2 SUSPENSION CULTURE

It has been shown that when sucrose is added externally to plant cell suspensions, (Figure 3.2) it typically hydrolyzes quickly to form glucose and fructose, which are then absorbed either passively or actively, the mechanism likely dependent on the species. The invertases that hydrolyze sucrose may be found in the plasmalemma, cell wall, or external media. Fructose and glucose absorption patterns are associated despite the fact that glucose is always eliminated from the media more quickly. But according to research on the effects of sucrose and glucose on culture growth and secondary metabolism, there are distinctions in the way their uptake and utilization are controlled. Reports of the use of other glucose-containing disaccharides should also be taken into account in this situation. For example, maltose and lactose can both be used as energy sources by specific cell lines, including Daucus carota and Medicago sativa. Although there are few reports on how these alternate carbon sources affect secondary metabolism, they generally seem to have a negative impact. Once glucose has entered the cytoplasm through the plasmalemma as glucose-6-phosphate, it can enter primary metabolism via hexokinase and proceed to glycolysis or the pentose phosphate pathway from there. The finding that some cell lines can thrive on fructose for just one growth cycle may support the idea that fructose uptake is mediated by glucose. However, certain cell lines are able to use solely fructose as their energy source and can persist indefinitely on a medium containing only fructose. Because of this, it is unclear how fructose enters plants' core metabolism. It appears that most of the time, enzymes have a limited ability to handle fructose, albeit it is debatable when this limiting phase kicks in.

There are two potential intermediates in relation to fructokinase(Kaplan et al., 2007). In addition to phosphorylating glucose, hexokinases also catalyze the synthesis of fructose 6-phosphate. Fructose-kinase is a different enzyme that might also be a possibility. Fructokinase catalyze the phosphorylation of fructose at carbon atom one to produce fructose-phosphate, which can subsequently be broken down by an enzyme called an aldolase to produce glyceraldehyde and dihydroxyacetone phosphate. The phosphorylated glyceraldehyde and DHAP can subsequently go through the pentose phosphate pathway or glycolysis.

The earliest stages of the majority of secondary biosynthetic pathways largely rely on the supply of reducing equivalents, which may be provided by the first oxidative steps of the pentose phosphate route/pathway, which produce NADPH, or by glycolysis, which produces NADH. The non-oxidative phase of the pentose phosphate pathway or glycolysis would be the point at which fructose-6-phosphate entered primary metabolism. Fructose-6-phosphate could enter glycolysis directly, via glucose-6-phosphate, the pentose phosphate pathway, or the starch synthesis pathway. In this regard, fructose-6-phosphate may be the favored intermediary for producing NADPH, which is a form of reducing power. It would be interesting to investigate whether a fructose-6-phosphate intermediate plays any role in plant cell metabolism.

3.3 MEDIA

Researchers have discovered that some cell lines can grow indefinitely on a medium that contains solely fructose. In an experiment to increase glucosidase levels, cell suspension cultures of potatoes (*Solanum tuberosum*) and evening primrose (*Oenothera biennis*) were cultivated on 2% fructose for 11 and 20 subcultures, respectively(Cresswell et al., 1989). Even while growth was largely unaffected, a comparison with cultures of the same cell lines grown in 2 percent sucrose revealed that there were considerable changes in the glucosidase activity. When 2 percent fructose was present compared to 2 percent sucrose, glucosidase activity significantly increased, demonstrating that the particular monosaccharide was the main effector. There was only a concentration-linked persistence of activity with glucosidase in evening primrose. These findings support claims that glucose inhibits glucosidase function. Although most glycosidases have reversible enzymatic activity, hydrolysis is typically the preferred response, and transferase enzymes are likely producing glycosides. The majority of plant heterosides that occur naturally are linked. The specificities of the glycosides vary, but the majority seem to be specific. They play roles in the hydrolysis of cyanogenic glycosides, flavonoid, steroid, or monoterpenoid glycosides, as well as the deactivation and activation of plant hormones like ABA, 2,4-D, and gibberellins. It is impossible to know the relevance of the proportionate increase in glucosidase activity in regard to secondary metabolism. Two key characteristics set glycosylated secondary products apart from aglycones: increased water solubility and lower chemical reactivity. They are therefore less cytotoxic and more readily collected in the vacuole.

The inference that could be made from the observation that fructose supply increased glucosidase activity is that any storage of secondary products as glycosides would be reduced, the kinetics favoring hydrolysis, especially when intracellular glucose concentrations are low, as would be expected in the later stages of culture growth. If the products are cytotoxic in their aglycone form, a decrease in the capacity of cells to store secondary products as glycosides is likely to be accompanied by a decrease in the possibility of secondary product accumulation. It has been extensively studied how secondary product-accumulating plant cells affect different carbon sources. Sucrose is typically the most effective sugar for secondary metabolic production; however, it is unclear exactly how this impact is transmitted. The scenario in sucrose medium, where fructose may be anticipated to be significantly more available following glucose use as carbon substrate, later in the linear development phase, is another issue that emerges from these data. The glycosylation of secondary products may consequently be suppressed during this period, despite the fact that they are typically observed to accumulate quickly. The digestion of nitrogen has a high energy requirement since nitrogen is supplied to plant cells as nitrate. Ammonia is a more energy-efficient form of nitrogen to provide than the more oxidized version, however the removal of NH_3 from culture medium does cause a sharp pH drop due to the typically ineffective buffering in tissue culture media. The NH_3 usually seems to be removed and used preferentially when it is present in tissue culture medium along with other more oxidized forms like NO^{3-}.

3.4 INGREDIENTS

Urea, glutamine, glutamate, alanine, or casein are additional N-sources that are utilized in tissue culture medium(Tsai & Saunders, 1999). It makes sense that organic nitrogen sources would meet both the need for nitrogen and the need for carbon. The observed effects of amino acids on nitrate reductase activity, culture development, and particularly secondary product production are varied. Amino acids have also been utilized to augment other N-sources. These organic nitrogen sources probably have a wide range of complicated consequences, including potential functions as carbon sources, precursors for secondary products, and regulatory molecules.

Figure 3.3. Claviceps. (Source: https://upload.wikimedia.org/wikipedia/commons/c/ca/Claviceps_purpurea.JPG, Creative Commons License).

The formation of Claviceps (Figure 3.3) and ergot alkaloids can be extrapolated to show that tryptophan addition to fermentation results in a reaction approaching the phenomena of enzyme induction. For instance, tryptophan mimics that are not converted into alkaloids have been found to increase alkaloid synthesis. Tryptophan was shown to promote secondary biosynthetic enzymes, such as dimethyallyl-tryptophan synthetase, on the ergot alkaloid route, and was needed to be introduced to the culture early during the development phase in order to stimulate alkaloid production. In plant cell culture systems, the impacts of secondary product precursors are typically varied and even conflicting. The precursor supply can be increased in the cell in two different ways: first, by adding to the medium, in which case the uptake mechanism may or may not be limiting; and second, by selecting for resistance to precursor analogues (or by treating with antimetabolites), in which case the intracellular level may be altered. There are several instances where the production of secondary metabolites is boosted using both of these strategies. However, it is frequently found that there are no increases in secondary product accumulation, and there are a number of reasons why this can be the case.

3.5 PRECURSORS

In an effort to produce cultures that over accumulate precursors like L-tryptophan and potentially over accumulate secondary metabolites, several cell lines have been chosen for their tolerance of antimetabolites (particularly amino acid analogues) (Mobin et al., 2015). This strategy has been successful in a few instances, although in many cases no overaccumulation of secondary metabolites is seen in such lines. The selected cell line may be tolerable due to mechanisms that do not provide an overabundance of the precursor amino acids, which is one possible fundamental rationale for this. Another possibility is that there are no additional substrates available, or that there are regulatory processes farther along the biosynthetic route that inhibit the conversion of the excess precursor into secondary metabolites. The culture cycle divides into three phases with arbitrary boundaries in terms of energy requirement. During the first phase, nitrogen and carbohydrates from the medium are rapidly absorbed, loading the carbon and nitrogen pools. A minimum amount of oxidation occurs, and monosaccharides are kept as starch. Cell division takes place during the second phase, with the main direction of flux via glucose-6-phosphate coming from the oxidation of incoming carbon. Cellular respiration is currently high, in carbon-limited media, some mobilization of carbohydrate reserves may also take place during this phase, but third phase is likely to be the most important one for this process because it involves the remobilization of these reserves and the generation of carbon skeletons for oxidation as well as the synthesis of structural elements and secondary metabolites. It is frequently asserted that the rate of secondary product accumulation and growth rate are inversely related.

3.6 GROWTH

Cell growth appears to be more or less unique in some culture methods in terms of secondary product level and biomass increase. The accumulation of monomeric indole alkaloids by C. roseus cell suspensions known as M_3, occurs primarily after cell division has stopped, pointing to a potential connection between the non-dividing state and the channeling of substrate into secondary biosynthesis(Verpoorte et al., 2004). The use of cell cycle inhibitors such colchicine, hydroxyurea, and trigonelline, which were discovered to generate transitory increases in the level of secondary products, connected with a suppression of cell division, provides additional support for this theory. When cultures of C. roseus cell cultures were fed with a variety of hormone therapies after growing on a hormone-free, 5 percent sucrose medium for 6 days, there was a linear and direct association between the dry mass/cell and callus/cell. The results suggest a relatively strict control on the amount of alkaloid that can be accumulated in a cell of a given mass, considering the variety of hormones used in the experiment. On this high carbohydrate medium, suspensions of all treatments contained cells with sizable plastids, which are likely starch grains. There is evidence that, at least in this instance, the flux of primary precursors into storage and structural biosynthesis versus secondary product biosynthesis may be tightly regulated, possibly by an enzyme, as it is unlikely that at this stage of the culture cycle, i.e., after 6 days, the maximum capacity for alkaloid storage/cell has been reached.

Undifferentiated cultures might not have the specialized membrane or organelle organization needed for the compartmentalization of metabolites and/or enzymes at the subcellular level. The frequent inability of such cultures to distinguish between, for instance, glandular structures for the accumulation of toxic secondary products or laticifers for the expression of partial biosynthetic pathways is likely to result in an overall energy flow toward primary metabolism or structural biosynthesis. This means that the absence of collecting compartments does not necessarily imply a complete loss of secondary biosynthetic potential. Instead, the enzymatic machinery may be intact and only need product removal to be fully functioning. No change in the levels or types of primary metabolites is likely to change the flux through the morphinan alkaloid pathway in cases where cellular differentiation is also a requirement for enzyme activity, as appears to be the case, for instance, in the morphinan alkaloid synthesis in *Papaver somniferum* cell cultures. Although it has been demonstrated that the addition of an accumulating compartment, such as a lipophilic phase, such as Miglyol (a liquid triglyceride), or ion exchange resins, can change not only the proportion of secondary product accumulating in the medium but also the overall level accumulated by the entire culture, this effect is not limited to the medium. Examples include the recent use of polymeric adsorbents to induce the generation of anthraquinone and nicotine by cell cultures of *Cinchona ledgeriana* and *Nicotiana rustica*, respectively(Malik et al., 2013). The protection of volatile or easily broken-down secondary metabolites from the aqueous, oxidative culture environment is one of the possible causes for the elevated levels.

One cell line from a cultivar showed comparable development patterns both with and without miglyol. But over the course of five subcultures, the second cell line produced from cultivar consistently demonstrated a higher dry weight yield on miglyol than in its absence. When these cells were stained with iodine to detect starch, a microscope examination showed that they were full with granules of starch. At the moment of each subculture, the Miglyol supplied to the media of this cell line took on a waxy look and contained partially solid lumps rather than the typical liquid spherical globules. The observations could be attributed to either a cell's stress reaction to the lipophilic phase or to the utilization of some of the miglyol by, for instance, triglyceride lipase breakdown, which releases glycerol and free fatty acids. Researchers are currently looking into how these cell lines use both glycerol and fatty acids. It has been noted that plant cell cultures employ glycerol as a carbon source. One of the distinctive functions performed by chloroplasts in plant cells is photoautotrophism. Due to the fact that cultured plant cells are grown heterotrophically in the dark alongside an organic carbon source, this special property is typically not expressed in them. To create photoautotrophic (PA) cells, which have formed chloroplasts and can grow in the light without the addition of carbon sources, we therefore need to choose cells that exhibit high pigment(Yang et al., 2000). The growth of chloroplasts in PA cells is still restricted. The mechanisms governing chloroplast development are poorly understood, particularly in mesophyll cells, PA cells, and other heterotrophic cells that are unable to generate completely differentiated chloroplasts. It is unclear whether these impairments are caused by genetic mutations, gene modifications, or other transcriptional or translational regulatory processes.

By using restriction endonuclease, some analytical findings on the nuclear genome and chloroplast DNA were achieved. Tobacco photomixotrophic (PM) cultures were kept alive in modified Linsmaier and Skoog liquid media containing 3 percent sucrose(Georgiev et al., 2009). PA cells were grown in the same media as those of PM cells without sucrose but with CO_2 enriched (1-2 percent) air. Tobacco (*Nicotiana tabacum cv. Samsun NN*) plants were cultivated in soil for two to three months in a greenhouse with natural light. Analysis showed that the smaller cell volume of mesophyll cells in green leaves compared to cultivated cells on a fresh weight basis was the primary cause of the increased ChI concentrations. The slight discrepancies between the two examined cell types could be explained by the fact that while the ChI level per chloroplast was nearly same, the number of chloroplasts per cell varied. Some findings suggest that changes in the amount of chloroplasts present in each cell across all cell types were the primary cause of changes in the ChI content of cells during growth. Photosynthetic activity measurements revealed that leaves and cultured cells produce chloroplasts differently. On a ChI basis, the highest PSI and PSII activity in cultivated cells were roughly half that of green leaves. Additionally, on a fresh weight basis, the PSI and PSII activities of leaves varied synchronously with the change in ChI content, although they did not closely track the changes in ChI content in cultured cells. PSI and PSII activity were fairly consistent in PA cells. Early in the exponential phase, PM cells displayed the highest levels of activity, which thereafter dropped.

Analysis of Thylakoid Membrane Polypeptides LDS (Lithium Dodesyl Sulphate)-PAGE revealed that the compositions of thylakoid polypeptides of cultured cells were almost identical to those of leaves at any growth phase. The variations in the quantity of thylakoid polypeptides, however, did not account for the changes in photosynthetic activity during growth. Total cellular proteins were examined using SDS-PAGE, and they revealed some similarities as well as some variations between mesophyll cells and cultivated green cells. PAGE was used to further separate the total cellular polypeptides. On polyvinyliden difluoride (PVOF) membranes, the separated proteins were electroblotted and then Coomassie Brilliant Blue was used to make them visible. We can identify the main proteins whose amounts varied among cell types by visually comparing the relative brightness of the protein spots. While cultivated cells accumulated protein PI-4 in proportionally higher levels than mesophyll cells, protein L_1 and L_2 were present in mesophyll cells in a substantially higher amount. Using the immunoblotting technique and a tobacco-specific antiserum, protein L_1 and L_2 were recognised as the large and small subunits of ribulose 1,5-bisphosphate carboxylase (RuBisCO), respectively. To determine the type of protein that is predominately accumulating in mesophyll and cultured cells, researchers sequenced several of the key proteins. To sequence the NH_2 terminal amino acid, PVOF membrane-binding proteins were fragmented and then directly exposed to Edmann degradation using the Applied Biosystems Model, a sequencer that uses the standard sequence software and has an online PTH analyser. P_3 failed to produce an amino acid sequence during the sequencing process, indicating that the NH2 terminus was blocked. For 35–38 cycles, the remaining PI, P2, and P4 sites could be sequenced. The recoveries of PTH, alanine, and valine from the PI protein at positions 1, 15, and 5, 13, respectively, were used to compute the average repetitive sequencing yield, which was calculated at more than 90%.

The information in the Protein Sequence Data Bank was compared to every protein sequence that was acquired. Researchers found that Nicotiana proteins share a considerable amount of homology with previously known proteins at their NH_2-terminal sequences. With the exception of three residues that were not found in sample, the NH_2-terminal amino acid sequence of protein PI, which has 37 amino acid residues, is similar to the NH_2-terminal amino acid sequence of osmotin. The P_4 residue-containing region also displayed significant similarity to the osmotin NH_2-terminal. NaCl-adapted cells have been shown to accumulate osmotin. A very high homology was found between the sequence-containing residue in P_2 and the NH_2 terminus of tobacco chitinase. While chitinase is known to be induced as a pathogenesis-related protein and also by the plant hormone ethylene, osmotin is known to be induced by NaCl stress. In addition to PA and PM cells, PI-P_4 proteins were also discovered in freshly formed suspension cells, in vitro-grown regenerating adventitious shoots, and newly established suspension cells. These findings indicated that in vitro cultivated cells are under extreme stress, however the causes of this stress are yet unknown. Chloroplasts can multiply semi-autonomously and have unique DNA. In recent years, it has been suggested that the production of albinos or the suppression of chloroplast development may be due to some deletion or methylation of chloroplast (plastid) DNA(G. Liu et al., 2020). In cultivated PM/PA cells, researchers investigated the potential that the alternation of nuclear DNA and chloroplast gene structures would have an impact on the gene expression in chloroplast function. The Hermann approach was used to isolate chloroplast DNA from PM cells and green leaves . After being digested with 12 restriction endonucleases and separated by agarose gel electrophoresis, the chloroplast DNAs of both cultivated cells and leaves were examined.

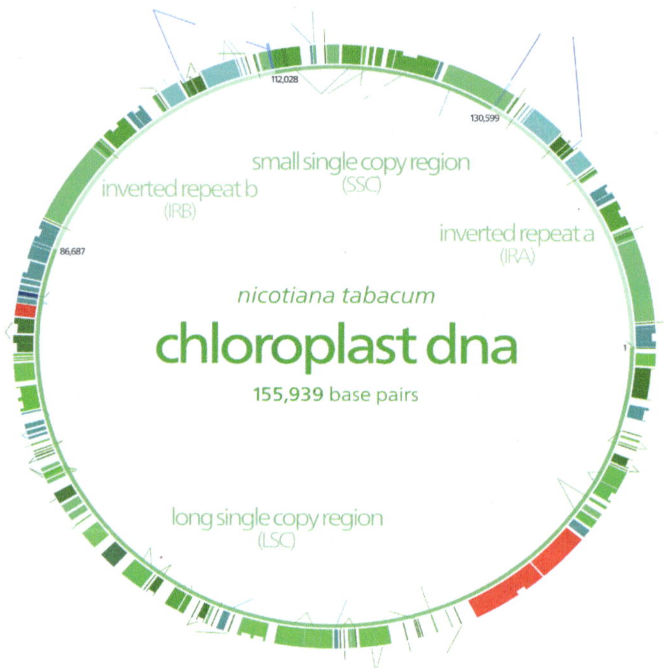

Figure 3.4. Chloroplast DNA. (Source: https://upload.wikimedia.org/wikipedia/commons/thumb/f/fc/Chloroplast_DNA_blank.svg/1200px-Chloroplast_DNA_blank.svg.png, Creative Commons License).

3.7 CHLOROPLASTS

DNA from chloroplasts (Figure 3.4) was discovered to have restriction fragment patterns that were identical to those of chloroplasts from leaves of the parent variety. A good agreement between them was also produced by Southern hybridization using a probe for a major subunit of the RuBisCo protein. The nucleotide sequence of the chloroplast psbA gene extracted from cultivated tobacco cells matched the parent variety exactly. By using a probe for rDNA and Southern hybridization following treatment with an endonuclease (Eco RI or Hind III), the structure of nuclear DNA was also examined. The findings showed that while the amount of rDNA in the nuclei of cultivated cells was only around one-fourth of that of mesophyll cells, the restriction pattern of the nuclear rDNAs of PM and mesophyll cells was identical. However, many tests revealed that when compared to mesophyll cells, the formation of chloroplasts in cultivated cells, even PA cells, was suppressed. The ChI content of cells on a fresh weight basis made the most difference. Except for one cell line of soybean, cultured green cells have not yet been reported to have a ChI content that is equal to that of mesophyll cells on a fresh weight basis. However, there was little variation in ChI concentration per chloroplast across the cells analyzed. The low ChI level of cultivated cells on a fresh weight basis was also slightly influenced by the amount of chloroplasts in each cell. It was evident that the difference in ChI content between mesophyll cells and cultured cells on a fresh weight basis was caused by the difference in cell volume between both. But the variations in the amount of chloroplasts present in each cell could account for the variations in the ChI concentration of cells during growth. Measuring photochemical reaction activity also identified some differences between mesophyll cells and cultivated green cells. As was previously documented in wheat and soybean leaves, these activities changed in green leaves at the same time as the ChI content on a fresh weight basis. The changes in photochemical reaction activity in cultivated cells did not closely correspond to the changes in ChI concentration.

The huge cell volume of cultivated cells, particularly at the high CO_2 compensation threshold, may be a factor in the low photosynthetic activity. The main cause of the high CO_2 compensation has been attributed to lower RuBisCo and carbonic anhydrase activities as well as high respiration activity. We suggest that the random arrangement of chloroplasts in cultivated cells also influences the activity of the photosynthetic process because it takes longer for CO_2 to transfer from the plasma membrane to the chloroplasts. Thylakoid membrane and stromal protein analyses revealed that, with the exception of some polypeptides, such as the subunit of CF I, LH C (I), LH C(II), and polypeptides of the reaction center of PSI and PSII and others, especially in PM cells, the composition of the chloroplasts of cultured cells was almost identical to that of mesophyll cells(Gounaris et al., 1986). All of the aforementioned analyses showed that even when grown under PA conditions, cultured green cells had less developed chloroplasts and reduced chloroplast activity. One explanation for the decrease in gene expression in cultivated cells is because nuclear or chloroplast genes that code for chloroplast proteins underwent mutation or modification. However, our research on nuclear and chloroplast genomes and DNA revealed that uncommon gene changes in PA/PM cultivated cells' nuclear or chloroplast genomes occurred. It was claimed that the number of copies of DNA in chloroplasts and mesophyll cells were comparable. It's interesting to note that mesophyll cells have more rDNA than PA cells did. Compared to newly produced cultured cells from the same tobacco type, PM cells develop more slowly. Chloroplasts may develop well with slow growth.

3.8 MECHANISMS

Various regulatory mechanisms on transcription and/or translation processes control the low gene expression of chloroplasts in cultivated cells. Total cellular protein analyses made it very evident that stressed plant cells in culture are present. Although it is unclear which stressors affect cultured cells severely, it is very likely that some forms of stress inhibit gene expression and chloroplast formation. For cultured cells' gene expression to be modified to boost productivity, a deeper knowledge of the regulatory systems that govern them is crucial. In addition to being an excellent model system for studying plant gene expression in vitro, photoautotrophic cells would also be helpful for selecting mutants with altered chloroplast functions. Recently, researchers were successful in isolating a mutant with an atrazine resistance gene psbA mutation. Due to the photomixotrophic cells' long-term maintenance after callus induction, whole plants could not be reproduced from resistant cell line. However, because to recent developments in protoplast fusion techniques, we can anticipate that this mutant gene will prove valuable for plant breeding. We can think that cultured cells provide a useful framework for selecting novel mutations in the nuclear and chloroplast genomes. For roughly 25 distinct plant species, with the exception of the Gramineae, photoautotrophic growth of in vitro cultivated plant cells has been achieved in the presence of CO_2-enriched air (1-2 percent, v /v) but in a sugar-free medium. Regarding its development requirements and its photosynthetic capabilities, including chloroplast differentiation, photoautotrophic plant cell suspension cultures reduce a complex plant organism to a single cell level. This makes it easier to research the physiology, biochemistry, and molecular biology of the green plant cell using photoautotrophic cell cultures.

The claim that photoautotrophically cultivated plant cells share many physiological and biochemical characteristics with mesophyll cells may now be confirmed in a number of ways. This is particularly true for membrane transport capabilities and photosynthetic capacities as well as for the composition of the chlorophylls, carotenoids, lipids, and their constituent fatty acids. However, there are some striking distinctions between whole plants and photoautotrophic plant cell cultures. Concerned are the high levels of malate produced by the PEP-carboxylase reaction, the requirement for high CO_2 concentrations (1-2 percent; v/v) for the induction and maintenance of photoautotrophic development, and finally the requirement for blue light-dependent chloroplast differentiation. To determine whether the blue light dependence of chloroplast differentiation is a phenomenon also seen in photoautotrophically cultured plant cells or whether it is only present in cells growing heterotrophically in a complex medium with the presence of sugar, phytohormones, and vitamins, photoautotrophic cell suspension cultures from C. rubrum have been used. Cell suspension cultures of C. rubrum were switched from white light (24 W m^{-2}) to either red light or blue light of identical energy flux rate. The cultures were then grown in a simple mineral salt medium with 2 percent (v/v) CO_2(Widholm, 1992). The photoautotrophic cells' exposure to red light prevented the synthesis of chlorophyll and hindered their ability to proliferate in photoautotrophic environments. Such red-light-grown cells had a chlorophyll concentration that was less than half that of blue-light-cultured cells. The influence of the two-light quality on the steady state level of the complementary mRNA sequences in photoautotrophic cells was investigated using cloned chloroplast genes that encode the major subunit of the ribulose-bisphosphate carboxylase (RuBP-carboxylase). The investigations revealed a similar rate of accumulation for the mRNAs of the light-harvesting chlorophyll a,b-protein (LHCP) and the small subunit of the RuBP-carboxylase, although only to a considerably lesser extent under red-light conditions. These findings show that blue light is primarily responsible for inducing and maintaining chloroplast differentiation.

Blue light regulates mRNA levels. Researchers were able to successfully isolate and purify a transcriptional complex (transcriptional active chromosome; TAC) from the chloroplasts of suspension grown cells from C. rubrum photoautotrophic cell cultures. By hybridizing with cloned chloroplast DNA fragments utilizing gel transfer techniques as well as dot hybridization, the plastid transcripts in the RNA produced by the TAC from chloroplasts of either blue- or red-light irradiated photoautotrophic cells were examined and quantified. When RNA was created from TAC in the chloroplasts of blue-light irradiated cells as opposed to red-light grown cells, the tagged transcripts of both TAC systems produced almost the same pattern of radioactive dots. The amount of gene-specific transcripts produced by TAC from blue-light irradiated cells was around 40% more than that of TAC from red-light growing cells, as evidenced by the $_{32}$P-label and consequently. It's possible that blue-light cells have substantially more RNA polymerase activity than red-light cells do. Red light irradiated photo autotrophic cell cultures exhibit much lower steady-state levels of mRNA encoding for prominent chloroplast proteins than blue light-treated cells, yet photoautotrophic growth continued after 14 days of red-light irradiation. It made sense to try to choose cell lines that could grow photoautotrophically indefinitely in the presence of red light. Researchers were successful in creating C. rubrum cell cultures, and they have been developing photoautotrophically for more than six months only when exposed to red or blue light with an equal photon flux density. Experimental results demonstrate unequivocally that the nutritional state of cultivated plant cells affects light-dependent chloroplast development.

3.9 LIGHT

Compared to red light of identical photon flux density, blue light significantly inhibits photoautotrophic development in cell suspension cultures of C. rubrum. This raises the question of whether growth inhibition is caused by a general decrease in photosynthetic CO_2 uptake and/or by significant changes in how specific photosynthetic products are formed. When measured on a chlorophyll basis, the photosynthetic CO_2 uptake of photoautotrophic cell cultures from C. rubrum is significantly reduced by blue light as opposed to red light. Only a few quantitative changes in the integration of photosynthetic carbon into the primary fractions were found after a thorough examination of the production of HC-labeled photosynthetic products following $^{14}CO_2$ photosynthesis. When compared to red light, blue light promoted ^{14}C-incorporation into amino acids and proteins at the expense of HC-labelling of organic acids and carbohydrates. The question of whether blue-light-induced growth inhibition in photoautotrophic cell cultures from C. rubrum is only caused by a reduction in photosynthetic CO_2 assimilation and product formation cannot be answered at the current stage. Alternatively, non-photosynthetic blue light effects on cell metabolism must also be taken into account. C. rubrum photoautotrophic cell suspension cultures indicate a C_3-type of photosynthesis since they primarily assimilate CO_2 through a C_1 to C_5 carboxylation sequence employing the RuBP-carboxylase. However, substantial levels of malate and, to a lesser extent, aspartate are also produced as a result of the PEP-high carboxylase's in vitro activity, particularly in actively dividing cells during the exponential development phase. Malate production rate was found to be distinctly pH-dependent(Rocha & Ting, 1971). The rate of HC-incorporation into malate and aspartate following short-term HCO_2 photosynthesis substantially doubles at the expense of HC-labelling of sugar phosphates if the pH of the incubation medium is raised from 4.5 to 6.0 or 7.0.

After selective decarboxylation of malate at the 4-C atom with the malic enzyme, the intramolecular pattern of HC-labelling of malate generated during short-term $^{14}CO_2$ photosynthesis was identified. Malate is created by the carboxylation of phosphoenolpyruvate, which is carried out by the PEP-carboxylase, and contains around 80%· of the total HC-radioactivity. In photoautotrophic cell cultures of C. rubrum, the PEP-carboxylase will significantly contribute to overall photosynthetic CO_2 assimilation under the right pH conditions. This implies that the extracellular pH level can control the expression of the C_4-type of carbon assimilation in the photoautotrophic cell. The concentration of dissolved inorganic carbon rises as the pH of the culture medium rises, mostly because more dissolved HCO is present at pH levels over 6. The maximum photosynthetic activities of the cells were seen in algae and isolated mesophyll cells. At lower or higher pH values, photosynthesis was inhibited. The observed increase in malate production following a pH shift from 4.5 to 6.0 or 7.0 in the extracellular space may be the result of a pH-dependent activation of PEP-carboxylase activity. This suggests that variations in the pH of the extracellular space will also affect the pH of the cytoplasm inside the cell. The intracellular (cytoplasmic, vacuolar) pH of the plant cells was measured using ^{31}P NMR spectroscopy as a non-destructive, in-vivo way to address this issue. Despite the fact that these experimental results must be viewed as preliminary, they will provide a preliminary estimate of the cytoplasmic pH in C. rubrum cells that have been cultivated in a photoautotrophic method. From the vacuole to the extracellular media, there appears to be a significant ion flux. For instance, if the incubation media was adjusted to pH 6.0 in the presence of MES-NaOH buffer or by NaOH, respectively, photoautotrophic cell suspensions acidified the culture medium for around 0.1 or 0.5 pH units per hour. Similar outcomes have been reported for heterotrophic plant cell cultures as well as mesophyll cells from higher plants.

3.10 PH

Although the extracellular pH value was raised from 4.5 to 6.0, the cytoplasmic pH in photoautotrophically cultivated cells of C. rubrum only changed for about 0.1–0.2 pH units; yet, this rise in pH may have significant stimulatory effects on the activity of the PEP-carboxylase. Researchers discovered that the PEP-carboxylase from photoautotrophic cell cultures of C. rubrum is substantially pH-dependent in its in vitro activity(Hüsemann et al., 1989). For instance, increasing the test medium's pH from 7.2 to 7.4 will cause a 1.8-fold increase in PEP-carboxylase activity. Short-term $^{14}CO_2$ photosynthesis at pH 6 should result in an increase in malate production, and this rise should be adequate to explain this increase. Cell separation of isolated protoplasts has demonstrated that the PEP-carboxylase in photoautotrophic cell cultures from C. rubrum is of cytosolic origin. It has been done using the silicone-oil centrifugation method. The activity of marker enzymes specific for chloroplasts (NADP-glyceraldehyde-3-phosphate dehydrogenase; NADP-GAPDH), mitochondria (fumarase), peroxisomes (catalase), and cytosol was used to track the separation of the chloroplasts from other cytoplasmic components (pyruvate-kinase).

Figure 3.5. *Morinda lucida*. (Source: https://upload.wikimedia.org/wikipedia/commons/thumb/1/12/P_Morc_D5915.JPG/800px-P_Morc_D5915.JPG, Creative Commons License).

According to the NADP-GAPDH activity, about 80% of the chloroplasts were separated into a pellet fraction within 8 seconds of centrifugation. This fraction was essentially free of soluble cytoplasmic components (5 percent pyruvate kinase activity), but it still contained more than 30% contamination by mitochondria. The modest size differences between the mitochondria and chloroplasts in actively developing cells as well as an attachment of the chloroplasts and mitochondria to multiorganelle complexes generated during centrifugation may contribute to the inadequate separation of the two organelles. However, PEP-cytosolic carboxylase's nature is confirmed by the subcellular fractionation of protoplasts using the silicone oil centrifugation method. It provides a preliminary estimate that the NAD-malic acid dehydrogenase (NAD-MDH) is limited to the mitochondria while the NADP-malic acid dehydrogenase (NADP-MDH) is localized in the chloroplasts and cytoplasm. An extra production of oxaloacetate and malate as intermediates of the citric acid cycle to support its synthetic function would be made possible by such an intracellular distribution of enzymes, supporting their potential involvement in metabolic events.

In cell suspension cultures of *Morinda lucida*, (Figure 3.5) anthraquinone and lipoquinone form. The lipoquinones are only present in photoautotrophic cultures, whereas anthraquinone pigments are restricted to heterotrophic cultures. Such studies reveal the role of chloroplasts as the location of lipoquinone formation and shed light on the regulatory parameters governing aromatic compound metabolism. *Ruta graveolens* heterotrophic and photosynthetically active cell cultures have markedly different volatile compositions. C_5-C_8 ketones, acetic acid n-butylester, and a string of C_4-C_9 primary and secondary alcohols are all produced by heterotrophic cells. Green cell cultures are distinguished by C_9-C_{13} methylketones and esters of 2- and 3-methylbutyric acids esterified with straight-chain or branched C_8-C_{11} alcohols. Low amounts of indole alkaloids, tryptamine, and tryptophan decarboxylase activity are found in photoautotrophic cell suspension cultures of *Catharanthus roseus*. The level of these elements dramatically rises when such cultures are converted to photomixotrophic and then heterotrophic cultures. Tryptophan decarboxylase and tryptamine buildup are very immediately induced when photoautotrophic cells are exposed to sucrose-rich conditions. These findings suggest that heterotrophic cells are the only ones capable of producing indole alkaloids and expressing the relevant biosynthetic enzymes.

The p-coumaric and ferulic acid glucose esters are produced by *Chenopodium rubrum* cell cultures. Because the biosynthesis is controlled by both the light-induced stimulation of phenylalanine ammonia lyase and by suitably high amounts of sucrose, these esters are particularly prevalent in photomixotrophic cultures. The cells' aromatic glucose esters are hardly noticeable in photoautotrophic conditions, and phenylalanine ammonia lyase exhibits very low enzyme activity. Sucrose can be used to titrate the induction of PAL and the buildup of cinnamic acid esters. The amounts of cinnamic acid esters and PAL activity are both relatively low below 0.5 percent of the carbon supply, but maximum expression of the enzyme has been observed between 0.5 and 1 percent sucrose. Researchers have found that the regulatory transition between the photoautotrophic and heterotrophic modes of nutrition occurs at a concentration of about 0.5 percent sucrose. The hypothesis that green cell cultures are appropriate models if chloroplast-derived products are being investigated and, furthermore, that sufficiently high concentrations of sucrose are available to sustain secondary product biosynthesis of a suitable rate, summarizes our current knowledge of secondary product formation in photosynthetically active plant cell cultures. Therefore, because photomixotrophic cell cultures have functioning chloroplasts and have enough of a metabolizable carbon source, they are the most appropriate for use in future research. The nutritional state (heterotrophic-photoautotrophic) of the cells determines their differential sensitivity to blue and red-light during chloroplast development. On a molecular biological level, studies on the mRNA steady-state concentrations and on the transcriptional activities of the cells are ongoing to explain this variation in light-dependent chloroplast differentiation. Further research is needed to determine whether this shift in the cells' sensitivity to blue/red light is also accompanied by a transition from a particular blue-light photoreceptor to phytochrome.

CHAPTER 4

Plant Tissue Culture and Biotechnology

CONTENTS

4.1 INTRODUCTION

The production and maintenance of plant cell, tissue, and organ cultures require an aseptic approach. The *in vitro* environment designed for the growth of plant material is ideal for the growth of bacteria as well. Usually, microorganisms invade the plant tissues. Furthermore, an infection can cross cultural barriers. To lessen the possibility of germs entering or remaining in the cultures, an aseptic method is adopted. Autoclaving (Figure 4.1) is the method most usually used to sterilize items made of heat-resistant materials.

Figure 4.1. Autoclaving. (Source: Shaik YP, Schuster J, Shaik A, Mohammed M, Katherapalli HR. Effect of Autoclave Pressure and Temperature on Consolidation of Layers and Mechanical Properties of Additively Manufactured (FDM) Products with PLA. *Journal of Manufacturing and Materials Processing*. 2021; 5(4):114. https://doi.org/10.3390/jmmp5040114, Creative Commons License)

Depending on the volume of each vessel and the number of vessels in the autoclave, the time spent in the autoclave will change. Organic compounds, including growth regulators, amino acids, vitamins, and antibiotics, may degrade during autoclaving. Filter membranes are required to sterilize these compounds. Numerous steps are often involved in the preparation of nutrition media that contain heat-labile components. The heat-stable components are dissolved in a sterile solution that has been autoclaved and then cooled to 45 to 50°C to sanitize it as usual. The solutions of the thermolabile components are sterilized. The sterile solutions are then combined in an aseptic environment to generate the whole medium. Contamination can be a particularly difficult problem when plant explant material is obtained from the field or greenhouse. The initial contamination is visible a few days after cultures are started. Bacteria induce turbidity in liquid cultures (Gamborg, 2002; P. P. Kumar & Loh, 2012).

4.2 CONTAMINATION

Bacterial contamination is frequently found long after cultures are started and is typically caused by bacteria. It appears that the bacteria are endogenously present in the original plant material and are not immediately harmful. The fact that latent contamination can spread freely between cultures makes it harmful. A poor sterile technique or unclean lab settings can also lead to contamination. With the right precautions, this form of contamination is largely avoidable.

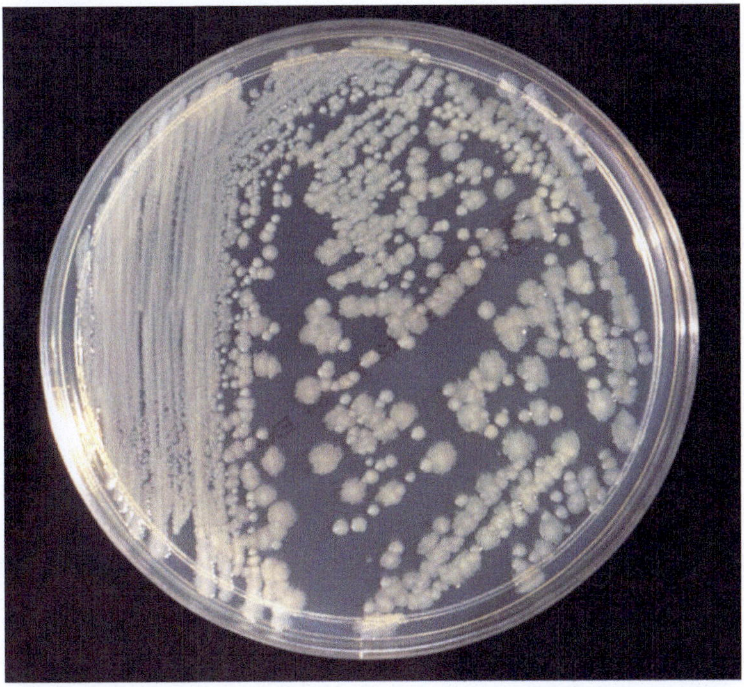

Figure 4.2. *Enterobacter ludwigii*. (Source:https://upload.wikimedia.org/wikipedia/commons/7/79/Enterobacter_cloacae_01.png, Creative Commons License)

The amount of contamination on explants may be reduced. Particularly in humid regions, plants grown in the field are often dirtier than those cultivated in a greenhouse or growth chamber. Explants can be obtained in an aseptic setting from seedlings made from sterilized seeds. The seeds will first be placed in the sterilizing solution for a period of two to three hours, and it is advised to use a magnetic stirrer. For each tissue and each plant species, the length of sterilization and the sterilization agent to be used must typically be determined empirically. Problems may arise from seeds that have an irregular seed coat or a hairy covering. Adding a few drops of a detergent-like Tween 20 or 80 might be beneficial. The seeds must be washed in water that has been autoclaved after being surface sterilized. The sterilized seeds are then either put into sterile agar media or sterilized wet filters in Petri plates for germination. The fewer seeds per vessel that are used, the higher is the risk of contamination.

Figure 4.3. *Jatropha curcas*. (Source:https://upload.wikimedia.org/wikipedia/commons/c/c6/Jatropha_curcas1_henning.jpg, Creative Commons License)

Here, explants are typically cut with the aid of a scalpel or a similar tool such as scissors that have been sterilized in a drying oven. Augmentin was determined to be the most suitable and effective antibiotic for preventing *Enterobacter ludwigii* (Figure 4.2) infection in *Jatropha curcas* (Figure 4.3) leaf explants. Without any additional contamination, the cultures can be grown for a prolonged period. Some findings demonstrate a considerable reduction in the number of microorganisms in tissue culture and plant cells. Vegelys™ can also be used as a surface sterilizing agent, and unlike commercially available treatments, it has no detrimental effects on seed germination. Fermenter or bioreactor cultures are subject to the same rules as those just mentioned. Although the bioreactor was initially designed to cultivate algae, it has been successfully utilized to cultivate cells of numerous higher plants. Before transferring cells, the bioreactor is placed in the culture chamber for three days to determine whether the autoclaving was successful. To avoid IAA from photo oxidizing, which will be visible in a few days if IAA is a component of the nutritive medium, the fermenter must be maintained in the dark. After that period, if the bioreactor is still sterile, the cell material is transferred using a sterile glass funnel and a silicon pipe with a diameter of 1 cm. To place the funnel in the inoculation cabinet's sterile air stream, the fermenter must be positioned in front of the laminar flow. The sterilization of the container and the nutrient media separately has also been used successfully as an alternative. The empty container was autoclaved for 35 minutes at 1 bar and 130°C to sterilize it. The nutrient medium was sterilized in the same manner as previously stated(Cassells, 1991).

4.3 BIOREACTOR

A bioreactor used to cultivate plant material needs to provide appropriate mixing while reducing hydrodynamic pressure and shearing stress. The most promising design to meet these requirements, airlift bioreactors has been the focus of extensive research and development since the 1970s. Batch cultures have been undertaken to set up continuous systems in bioreactors. The cells are fixed on a stationary carrier in this instance, similar to animal cell cultures. Animal cells can adhere themselves to a glass surface or to synthetic materials like Sephadex, while plant cells find it difficult to do so, perhaps because of the tough cell wall. The capture of the cells inside the carrier material offers a solution to this challenge. Some polymers, including agar, agarose, polyacrylamide, and gelatin, have been investigated since calcium alginate was first utilized as a carrier. Pure synthetic materials including polyurethane, nylon, and polyphenyl oxide were also tried. All of these have benefits and drawbacks, but polyurethane is frequently chosen. The capture of the cells is caused by a passive penetration of the carrier material, which has a substantial interior volume. Cells continue to divide and expand in the pores when the carrier is immersed in the cell suspension, invading the whole interior volume. Since no additional chemicals are needed to attach the cells to the carrier, there are currently no known detrimental effects on the health and metabolism of the cells using this method. In the typical nutritional media, polyurethane is stable, even over long testing times. These polyurethane-fixed cells can be moved to a flatbed container or a column where they are continuously submerged in nourishing media(Wagner & Vogelmann, 1977).

Chemicals from the primary or secondary metabolism of plant cells are produced and can also be extruded into the medium. This provides the opportunity to alter the nutrient medium's composition to control cell growth at specific stages of culture. For a cell culture system to be successful, nutrient media are crucial. The majority of higher plants can grow autotrophically in light, under typical air conditions, with a sufficient supply of water and mineral nutrients, but not all plant organs and tissues can. For instance, for roots or growing seeds to survive, function, and grow, they need to import assimilates from shoot tissue or phytohormones produced in other distant tissue. This condition is also typical when cells from different cell culture systems are removed from the whole plant. The nutritional medium serves as a replacement for substances that would otherwise need to be imported and obtained from other portions of the entire plant that have different metabolic characteristics. Although some cell culture systems have been reported to grow photoautotrophically in an inorganic nutrient medium, by far the majority of culture systems are either heterotrophic or, in the light after the development of chloroplasts. For the cultures, addition of carbohydrates to the medium is necessary to meet the requirements of energy, as well as carbon, oxygen, and hydrogen as raw materials for synthesis. Typically, mono or disaccharides are provided for this reason. Various amounts of sucrose are used in media. Monosaccharides can be used to achieve good growth, and many additional materials, some of which are rather unusual, are also used. In general, there isn't the best carbohydrate for all plant cell cultures; the one that will be selected as an addition to the nutrient medium always relies on the tissue, the target of the study, and the type of plant. Initial investigations must be used to determine this.

Typically, sucrose concentrations range between 2 and 3 percent. The combination of mineral salts, which is a second essential element of a nutrient medium, has experienced significant alterations. The majority of modern media have an enhanced phosphorus concentration, which is a significant deviation from this. This concentration could be boosted by a factor of roughly 10 by adding coconut milk to the medium. The majority of the other mineral components are now present in higher concentrations as well. The addition of copper and molybdenum has broadened the category of micronutrients. Nitrate is the only inorganic nitrogen source in White's medium. For plants grown in light, it takes 8–10 days to establish enough functional chloroplasts that can serve as a reliable system to reduce nitrate. Only glycine is added to White's basal medium as a source of reduced nitrogen. The addition of casein hydrolysate to nutritional media, which contains several amino acids, has a similar function but is more effective.

Figure 4.4. MS medium. (Source:https://upload.wikimedia.org/wikipedia/commons/2/2b/Mammillaria_sp._%28006%29.jpg, Creative Commons License)

According to the Murashige and Skoog nutrient medium (MS medium), (Figure 4.4) ammonia is also provided in media as a source of reduced nitrogen in the form of different salts. The amount of phytohormones or other growth factors that are needed, as well as their impact on the growth and development of cultured cells, are largely determined by the plant species and variety, the tissue utilized for ex-plantation, and the purpose of the study or other use of the cultures.

4.4 PHYTOHORMONES

If original explants also have meristematic areas, then significant growth can already be generated in a hormone-free medium, and this growth is typically amplified by the addition of appropriate auxins. In contrast, quiescent tissue, like the secondary phloem of the carrot root, is the source of explants, development without hormone stimulation is very weak and consists primarily of cell expansion.

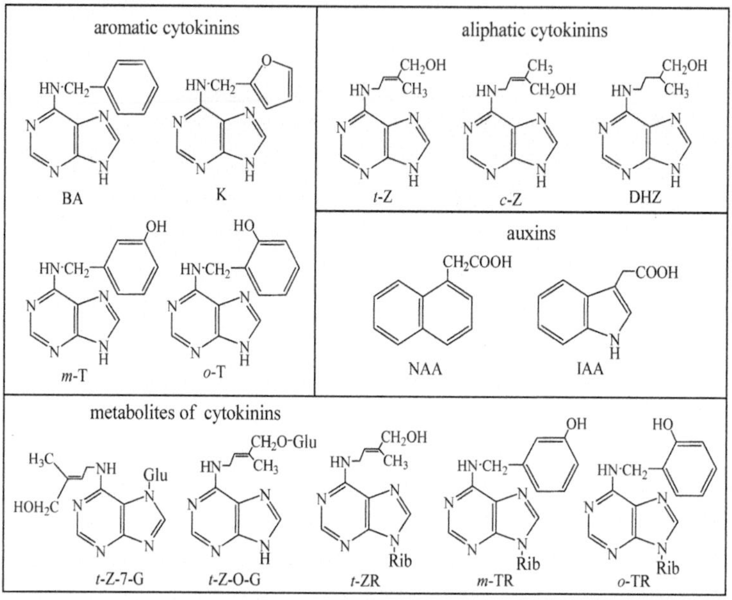

Figure 4.5. Phytohormones. (Source: Diopan, V., Adam, V., Havel, L., & Kizek, R. (2009). Phytohormones as Important Biologically Active Molecules – Their Simple Simultaneous Detection. *Molecules*, *14*(5), 1825–1839. https://doi.org/10.3390/molecules14051825, Creative Commons License)

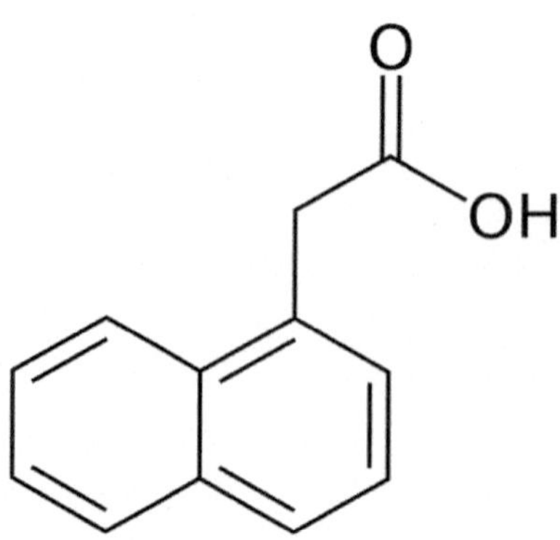

Figure 4.6. Naphthylacetic acid. (Source:https://upload.wikimedia.org/wikipedia/commons/thumb/6/63/Kwas_naftylooctowy.svg/1200px-Kwas_naftylooctowy.svg.png.jpg, Creative Commons License)

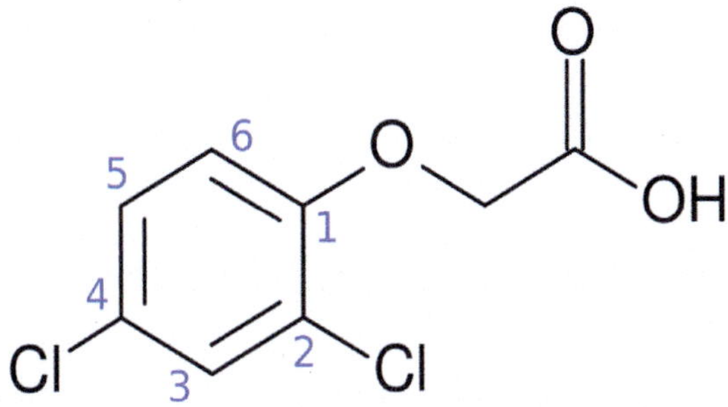

Figure 4.7. 2,4- dichlorophenoxy acetic acid. (Source:https://upload.wikimedia.org/wikipedia/commons/thumb/f/f2/2%2C4-Dichlorophenoxyacetic_acid_structure_numbered.svg/1200px-2%2C4-Dichlorophenoxyacetic_acid_structure_numbered.svg.png, Creative Commons License)

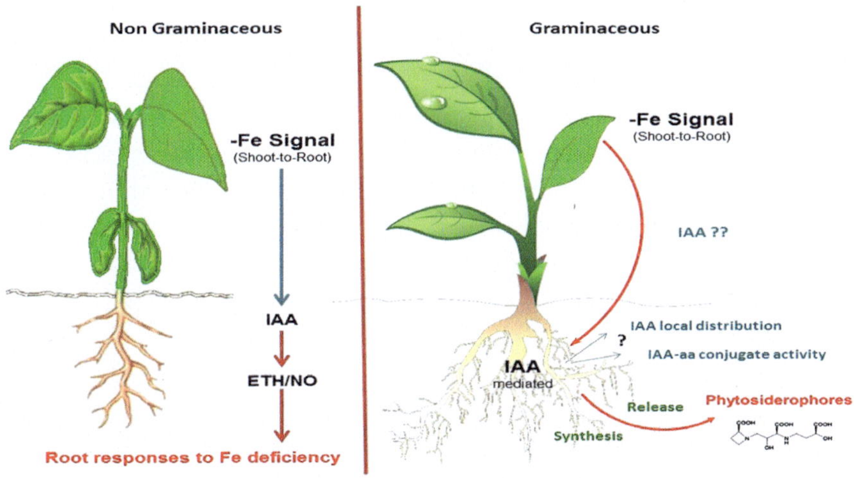

Figure 4.8. Indole acetic acid. (Source: Garnica, M. & Bacaicoa, Eva & Mora, V. & Francisco, Sara & Baigorri, Roberto & Zamarreño, Ángel & Garcia-Mina, Jose. (2018). Shoot iron status and auxin are involved in iron deficiency-induced phytosiderophores release in wheat. BMC Plant Biology. 18. 10.1186/s12870-018-1324-3., Creative Commons License)

Auxins (Figure 4.5) are present in the majority of nutritional media, typically in the form of naphthylacetic acid (NAA) (Figure 4.6) or 2,4-dichlorophenoxy acetic acid (2.4D) (Figure 4.7) and other auxins also exist that aid in the tissue culture. The natural auxin indole acetic acid (IAA) (Figure 4.8) is also occasionally added as a supplement. Chemical and metabolic resistance to breakdown or inactivation distinguishes these three auxins from one another. IAA quickly leaves the body, primarily by photooxidation but partly due to metabolic breakdown. NAA has greater stability, with 2.4D showing the most stability. In rapeseed cultures, differences in stability are correlated with the time needed to promote rhizogenesis, with IAA treatments showing the earliest development of roots. Auxins primarily promote cell division. The metabolically stable 2,4D auxin is the auxin of choice if a high cell division activity is to be maintained for extended periods, which typically precludes the differentiation of cultures. IAA and NAA are better dietary supplements if the goal of the experiment is to start differentiation processes, which typically call for brief periods of cell division. Doubling the auxin concentration will typically help to prevent the emergence of adventitious roots, though, which is frequently undesirable in basic cultures(Phillips & Garda, 2019). These are merely some basic observations, though, and since more accurate information about the plant hormonal system is not yet accessible, each culture system's auxin processing needs to be determined empirically.

By simultaneously adding a cytokinin to the nutritional medium, it is possible to further promote cell division activity, especially when using auxins. Kinetin, a synthetic cytokinin, is frequently employed in extremely low quantities (0.1 ppm). Higher levels can be extremely harmful. Natural cytokinins like zeatin or 2-isopentenyladenine (2-iP) and the synthetic 6-benzyladenine are also used in various culture systems (BA). Numerous cellular membranes contain inositol, which is crucial to the function of cell signaling systems. Although inositol is not a hormone due to the relatively high inositol concentrations employed in many nutritional media, significant responses similar to those linked to hormones are frequently seen. However, the concentration is insufficient to classify it as a nutrient or as an energy source. However, it was discovered in the 1960s that inositol might occasionally substitute IAA in terms of functionality. Since then, the existence of an IAA-inositol conjugate has been isolated. One potential function might be the building of a reservoir of such conjugates to prevent the breakdown of IAA, or the formation of such conjugates could be a method of inactivating it. Most of the time, m-inositol enhances the effects of cytokinins and IAA. Sterile filtration is performed at least for these ingredients if thermolabile components used in nutritional media run the risk of being altered or destroyed by autoclaving for the sterilization of the medium. For instance, any component containing radioactive isotopes is also sterilized in the lab.

4.5 NUTRITIONAL MEDIUM

The nutritional medium or other substances are passed through a bacterial filter with a pore size of 0.2 microns for sterile filtration. There is a wide variety of equipment for this technique available on the market. The easiest and least expensive method is to use a syringe with a filter adapter; however, this method works best with small amounts of liquid. It is frequently desirable to keep some cell strains alive for extended periods of time, sometimes even up to several years. This is especially true when substantial research is needed to connect several metabolic domains, which calls for the utilization of the same genome. It is necessary to regularly grow subcultures in order to preserve such cell strains or cell lines. This is necessary because of the cultures' expansion, their consumption of the nutrient medium's constituent parts, and the buildup of their excretions or dead cells. Depending on the severity of these conditions, subcultures can form at intervals as long as months or even years. For instance, a subculture required to be established every two or three days for *Phalaenopsis* cultures that produced large amounts of polyphenols discharged to the medium. However, a subculture interval of 6–8 weeks was sufficient for *Arachis* cell cultures that grew slowly and photoautotrophically. On agar, subcultures are typically immobile. A technique that works for many plant species' subcultures entails the aseptic transfer of callus fragments that are actively growing (10–15 per vessel) and have a diameter of around 5 mm to an Erlenmeyer flask (120 ml) containing 15 ml of nutritional solution. The subculture gap typically lasts four weeks(Leifert et al., 1995).

The cytogenetic instability of the cell material is an essential drawback. Other adverse effects include modifications to metabolic systems or even changes to the genome organization. These programs typically need a lot of work and storage. Changing the culture conditions, such as lowering the temperature of the culture room, limiting the nutrients, and reducing light intensity, can help create storage with frequent subcultures. The temperature can be lowered to almost zero degrees Celsius for cold-tolerant species. Cryopreservation was modified for plant cell cultures to avoid these issues, particularly during long-term storage. Cell suspensions or somatic embryos of cell cultures can be preserved with this method, which was first established for animal systems; following thawing, these can be revived with up to 80% success. Cryopreserved somatic embryos and cell suspensions can be stored for a number of years. The temperature for storage is -196 degrees Celsius, which is the point at which metabolism and cell division stop. Coconut milk no longer holds the same level of importance it once had for tissue culture systems. When all other nutrient media have failed to encourage the growth of explants, coconut milk can frequently be used as a supplement. Cultured carrot root explants in NL3 had a fresh weight that suggests an initial lag period of 5–6 days, an exponential phase of 2–3 weeks, and finally a stationary phase. The potential of the cells to develop depends on an increase in cell quantity, i.e., on cell division activity, during the exponential growth phase. As a result, the average cell size is smaller than it was in the original explants.

The cells of the cultured explant later increase in average size due to a slowing down of cell division activity. The average cell size has returned to that of the initial explants. Explants from diverse carrot roots exhibit a wide range in length. Furthermore, although being cultivated under the same conditions, cultures of diverse origins exhibit noticeably distinct patterns of cell division activity during the exponential phase. The carrot callus may appear to be morphologically unstructured, yet even a naked eye examination of cuttings reveals the existence of anatomical layers. A layer of two or three rather large cells (in width) can be seen on the perimeter, followed by a broad layer of smaller cells. Remains of the original explant are frequently found in the center itself. The callus surface appears to be compact and is made up of morphologically identical cells. However, the surface of Datura and Arachis cultures is rather loosely organized, and deep channels can be seen piercing the cultured explant through multiple layers of cells. When examining the Datura culture, it is easy to see that the cell forms are clearly distinct. The callus on the carrot root has a unique surface. From the sixth day on, newly developing cells from the explant may be seen pushing aside the remnants of cells that were removed when the original explants were obtained. This suggests that two or three cell layers beneath the surface of the explants, cell division has already begun.

4.6 SUSPENSION CULTURE

These are tiny cell clusters that are typical of actively dividing cell suspensions (Figure 4.9) and include roughly 100 cells each. Very small cells may be seen in the core of these meristematic meshes, while cell size rises toward the periphery. Although small, newly formed tracheids occasionally occur, it is safe to infer that the small layer of larger cells at the periphery of the explants should be from the periphery of meristematic mesh near the surface of the explants where cell division activity is reduced, and the age of the individual cell increases. The layers of larger cells in the core of the explants exhibit the same characteristics.

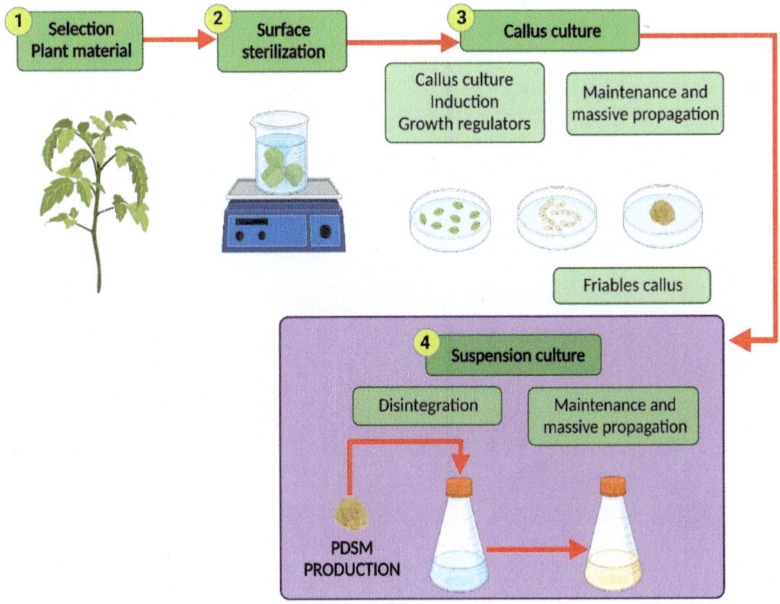

Figure 4.9. Cell suspensions. (Source: Motolinia Alcántara, Elizabeth & Castillo-Araiza, Carlos & Rodríguez-Monroy, Mario & Roman-Guerrero, Angelica & Cruz-Sosa, Francisco. (2021). Engineering Considerations to Produce Bioactive Compounds from Plant Cell Suspension Culture in Bioreactors. Plants. 10. 2762. 10.3390/plants10122762., Creative Commons License)

The center of these meristematic networks should be where cells in a developing callus with strong cell division activity propagate the fastest. It has already been mentioned that 24 hours following the start of culture, IAA synthesis and, to a lesser extent, 2iP synthesis take place. It appears that the explants are able to create an endogenous hormone system quickly after the start of culture(Xu et al., 2011). In conjunction with the hormones given by the nutritional medium, this natural hormonal system has the ability to define the direction of future growth. These exogenous phytohormones appear to be maintaining the ongoing cell division activity. Cell division halts in explants cultivated in a hormone-free media after roughly two rounds. As a result, hormone-supplemented treatments are able to create meristematic networks with meristematic cells that hormone-free treatments are unable to, and these nests are likely able to develop the endogenous hormonal system necessary for the continued development of the cultured explants. It is vital to know that the results of physiological or biochemical studies of cell populations, such as callus cultures, are averages of the responses of all the cells in the setup. Strong differences between individual cells must be taken into account, as was demonstrated by the cytological and morphological differences discovered by microscopic inspection.

Figure 4.10. *Welwitschia mirabilis*. (Source: By Thomas Schoch - own work at http://www.retas.de/thomas/travel/namibia2003/index.html, CC BY-SA 3.0, https://commons.wikimedia.org/w/index.php?curid=650590, Creative Commons License)

Another factor is cellular performance disruptions if environmental changes result from the employment of experimental equipment during such investigations. The differentiation of the nutritional systems also takes place as a result of these changes in differentiation during culture. Heterotrophic nutrition predominates for the first 8–10 days after culture initiation, using sucrose as a source of carbon and energy. Reduced nitrogen is provided by casein hydrolysate, which is a component of the nutritional medium and contains amino acids. Nitrate can be used as a source of nitrogen once the photosynthetic system is established. As carbon and energy are produced by photosynthesis, a mixotrophic diet is developed starting around the tenth day. The cultures can now transfer into a medium devoid of sugar and continue to thrive through photosynthesis. The cultures have finished using up the casein hydrolysate and sucrose that were initially given to them at time zero, and finally autotrophic feeding has taken predominance. This change also marks the beginning of the stationary growth phase. Transferring the explants at this point into a new media that has been supplemented with sucrose and casein hydrolysate starts a new round of intense cell division, and annual rings are beginning to form. Even with a plentiful supply of nutrients in the medium and explants cultivated on IAA and inositol but without kinetin, a transition into the stationary phase happens between days 10 and 12 of culture. The development of adventitious roots on the explants can typically be seen when the kinetin-free cultures enter the stationary phase. Shoots appear later, followed by the development of adventitious roots when using explants of rapeseed, Datura, and other plant species. The structures that finally form when roots and shoots connect together can be removed and then put into the soil to be further grown.

Although rare, *Welwitschia mirabilis* (Figure 4.10) is a crucial member of the phylogeny of primitive gymnosperms. It belongs to the monogenetic Welwitschiaceae family. Despite its slow growth, it has proven to be a useful experimental tool for studying anti-aging genes. Although structured, shiny, smooth, spherical formations were produced and the induced callus grew more quickly, no shoots could emerge from the callus. The majority of cell suspension cultures develop from callus cultures primarily as a result of mechanical impact in liquid medium. In stationary cultures on agar, a suspension is frequently made by squeezing a scalpel or using a sterile glass rod. Particularly with 2,4D in an agar medium, a loosely linked cell population can be removed with a scalpel on the other side of the agar. Using ammonia as a nitrogen source frequently results in an improvement, perhaps because cells have to expel protons in order to take it up. In a liquid nutritional media, callus cultures are typically agitated, and after 10 to 14 days, this mechanical impact causes the production of cell suspensions made up of cells from the periphery. Such a solution also contains material from dead or decomposing cells in addition to healthy cells that are still growing. Even now, there are varying opinions on what constitutes a cell suspension. Using a sharp scalpel, the highest (youngest) internode is explanted. To prevent the agent used for surface sterilization from entering the cut ends during sterilization, they are briefly submerged in liquid paraffin before being immersed for 5–6 minutes in the hypochlorite solution. Internode segments that are 1-2 cm long are cleaned four to five times with sterilized distilled water while being handled in a laminar flow.

The tissue is then divided into pieces that are about 1 mm thick and the paraffin cover and epidermis are removed with the aid of a sterile blade. These discs are split in half, and the callus cultures created by those halves are used. These segments weigh about 7-8 mg and have roughly 30,000 cells each, making them larger than the segments required to produce carrot cultures. There will be more explants available if the disc diameter is even larger. Within three weeks of culture, a callus that is highly proliferative and from which peripheral cells are easily removable with the use of a knife forms. For growth at 27 °C with a 12 h light/dark period, this cell material is transferred to the MS medium with agar supplemented with kinetin. The successive subcultures are started every two weeks after the first two subcultures, which are separated by three to four weeks. Only peripheral cell material from freshly formed callus fragments is used for subculturing as well. When there is enough cell material produced, the clusters are put into Erlenmeyer flasks on a shaker with a liquid medium of the same composition. Typically, a cell suspension consists of three fractions: free single cells of different shapes; cell aggregates made up of 10 or more cells; and, lastly, cell groups with a threadlike appearance. These fractions can be separated using the right sieving methods. Cell size rises toward the periphery in both, with extremely small cells visible in the center. In the middle of these structures, there is the most cell division activity.

The outermost cells of the cell aggregates are mechanically removed by the agitation, and these cells subsequently reflect the proportion of free single cells. These cells need to be more mature and inactive in terms of cell division. Some of these cells do, however, retain or re-induce the ability to divide. The thread-like structure is the outcome of three cell divisions. The four core cells showing chlorophyll accumulation would be the youngest cells generated from the recent cell division. One terminal cell divides into a tracheid-like structure, another accumulates anthocyanin, and one differentiates into a tracheid-like structure. The structural variations between the two terminal cells indicate an uneven initial cell division with irregular cytoplasm distribution. For both cells, the nutritional medium can be thought of as being the same. Only the cells in the aggregates had nearly homogeneous cytogenetics. The core of the meristematic networks with active cell division is where the lowest C-values of a ploidy level can also be found. In both instances, it was discovered that these minute cells in haploid cultures had DNA contents that were either twice as high in G_2-phase cells or nearly identical to microspores of the same species (G_1-phase cells). The DNA content of haploid G_1-phase cells in haploid cultures was either doubled in haploid cultures or was four times higher in haploid G_2-phase cells. A wide range of C-values was found in older cells found in the callus material between meristematic networks, which would be comparable to the percentage of free cells in the suspension. It appears that cytogenetic stability is correlated with the age of the cells, or the time since the previous cell division. Increased cell division activity caused by kinetin leads to higher cytogenetic stability and homogeneity of the cell population.

4.7 FERMENTER CULTURES

Fermenter cultures used for biotechnological applications can also be divided into a propagation phase and a production phase. Large fermenters are employed; however, to first make cell suspensions for inoculation, smaller laboratory fermenters are originally used. The cell suspension is often transferred from the smaller to the larger fermenter along with some nutritional media. It is standard procedure to remove some of the cell material from a semi-continuous culture at regular intervals for processing and to add fresh nutritional medium. Plant cell suspensions are already being cultivated today in fermenters with a volume of thousands of liters, for example, to manufacture Shikonin derivatives utilizing cultures of *Lithospermum officinale*(Figure 4.11).

Figure 4.11. *Lithospermum officinale*. (Source:https://upload.wikimedia.org/wikipedia/commons/4/4c/Lithospermum_officinale.jpeg, Creative Commons License)

Somatic embryogenesis has been used to reproduce cells in a fermenter. Subcultures must be produced regularly, typically every 1 or 2 weeks, and with a dilution of 1:5 after 1 week and 1:10 after 2 weeks with the fresh nutritional medium in order to retain cell strains in a healthy condition for extended periods. For each distinct strain, the best dilution and subculture frequency must be determined. Cryopreservation is frequently employed to retain cell suspensions (DiCosmo & Misawa, 1995).

Figure 4.12. *Solanum lycopersicum*. (Source: Bělonožníková K, Hýsková V, Vašková M, Křížek T, Čokrtová K, Vaněk T, Halířová L, Chudý M, Žufić A, Ryšlavá H. Seed Protection of *Solanum lycopersicum* with *Pythium oligandrum* against *Alternaria brassicicola* and *Verticillium albo-atrum*. *Microorganisms*. 2022; 10(7):1348. https://doi.org/10.3390/microorganisms10071348, Creative Commons License)

Plant cell walls can be hydrolyzed to release their macromolecular constituents, resulting in the production of naked cells known as protoplasts. These protoplasts are healthy and capable of surviving in isotonic media. Since the 1960s, protoplasts have also been employed to address many other fundamental issues, such as practical plant breeding issues. The production of plant hybrids devoid of cross-pollination in order to have plant material with traits unique to both parents has long been an aim of plant breeders. A histological examination of the shoot apex demonstrates that the hybrids produced are chimeras with rather fruit and leaf morphologies. In the 1970s, hybrids of tomatoes and potatoes created by fusing protoplasts of the two species were reported as a result of the first successful fusion experiments with protoplasts of distinct species. Researchers fused mesophyll protoplasts of *Solanum lycopersicum* (Figure 4.12) with callus protoplasts of *Solanum tuberosum*, however, the resulting plants were sterile.

Despite the nuclei fusion, restriction tests of chloroplast DNA and RuBisCO characterisation of both parents as well as the hybrid by electrophoresis clearly demonstrated the existence of two types of hybrids. The number of chromosomes was still greater than either of the parents. It appeared that one type of hybrid exclusively had potato plastids, whereas the other type contained tomato plastids. Microscopically identifying the fusion of potato protoplasts with protoplasts of light-green tomato plants carrying a genetically damaged chlorophyll system led to the identification of a successful fusion. Chlorophyll was generated when the potato cells were moved from darkness to light. Only accidental roots were regenerated by callus cultures of the tomato parent. The hybrids of the protoplast fusion trials had normal levels of chlorophyll and were either of potato origin or hybrids produced by the fusion of two protoplasts. It was feasible to differentiate between potatoes and the hybrid based on a variety of physical characteristics. Although these hybrids couldn't be employed in actual plant breeding operations, these early trials demonstrated the feasibility of producing crosses between diverse species. The original goal of protoplast cultures was to create new genomes with traits that none of the parents have. Protoplasts are frequently used to grow plants by somatic embryogenesis that can be used in plant breeding projects as well as to receive foreign genetic material. Direct insertion into rice protoplasts yields the best results and the highest efficiency is attained with the use of biolistics-based or *Agrobacterium*-mediated approaches for transferring foreign genetic material to competent cells (Figure 4.13).

The tissue used to produce protoplasts is sterilized by first exposing it to 70 percent ethanol for one minute, then immersing it in a 0.6 percent hypochlorite solution for 20 minutes. The leaf material is then transferred for 15 minutes to the nutritional medium utilized later for the development of the protoplasts, free of the organic components, and washed four times. 0.4 M mannitol is used to plasmolyze the plasmalemma and separate it from the cell wall. Following this pre-incubation phase, the leaves are divided into pieces that are around 0.5 mm long.

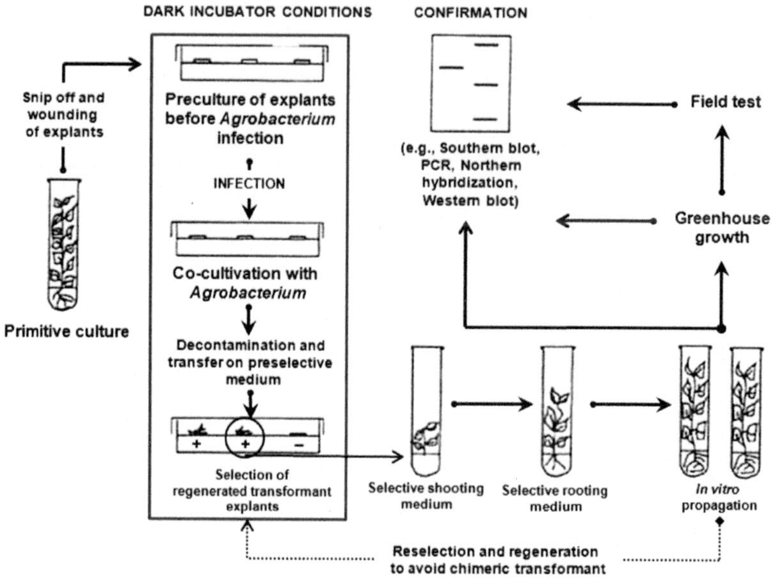

Figure 4.13. *Agrobacterium*-mediated approach. (Source: Movahedi A, Zhang J, Amirian R, Zhuge Q. An Efficient *Agrobacterium*-Mediated Transformation System for Poplar. *International Journal of Molecular Sciences*. 2014; 15(6):10780-10793. https://doi.org/10.3390/ijms150610780, Creative Commons License)

The enzyme solution is first sterilized by passing it through a membrane filter. To prevent illumination, the Petri dishes are covered with aluminum foil and sealed with Parafilm. The leaf material is placed in the enzyme solution for 6 hours at 28 °C in complete darkness. The individual protoplasts are gently shaken after incubation. The residual leaf material is eliminated by running through glass wool or a glass filter. After that, the protoplasts are gently centrifuged at around 100 g with a hand centrifuge to separate them from the incubation medium. The sedimented material containing the protoplasts is then washed four times with the nutrient medium that will be used to cultivate the protoplasts. The leaves that were 6 weeks old were used with the greatest success in tests to produce protoplasts of different Brassicaceae species. The ideal lighting time can be 10 hours. The cells may build up enough energy from photosynthesis during this period of light to endure the strain of transforming into protoplasts. At the start of culture, as demonstrated for cultured cells of tobacco and *Arabidopsis thaliana*, it was possible to successfully substitute the enzymes in the solution with the culture supernatant from *Clostridium cellulovorans* (Figure 4.14).

Figure 4.14. *Clostridium cellulovorans.*

(a) Cell surface of *C. cellulovorans* grown with cellobiose and (b) cell surface of *C. cellulovorans* grown with cellulose. White bar indicates 2 μm. (Source: Tomita H, Tamaru Y. The Second-Generation Biomethane from Mandarin Orange Peel under Cocultivation with Methanogens and the Armed *Clostridium cellulovorans*. *Fermentation*. 2019; 5(4):95. https://doi.org/10.3390/fermentation5040095, Creative Commons License)

Culture is carried out at 500 lux for the first 40 hours following protoplast formation, then at 2000 lux for the next 5 days: 26 to 28 °C is maintained, with 12 hours of light. After 10 days, cell aggregates are placed on agar. The free-floating protoplasts and tissue breakdown can be seen about 3 hours after the incubation began.

In the spherical protoplasts, chloroplasts first assemble along the cell edge and then, later, or approximately 30-35 hours after the experiment starts, they build up around the nucleus. These organelles frequently have a brownish color. The beginning of cell wall remodeling can be seen when Calcofluor White, a stain specific to cell wall material, is used. At the same time, the initially spherical protoplasts transform into oblong shapes, and 100 hours after isolation, the first cell divisions can be seen starting. It appears that starting cell division requires regeneration of the cell wall; a logical explanation for this occurrence has not yet been found.

The primary goal of protoplast fusion has been to join two genomes that cannot be joined through pollination. This goal can be more precisely attained because of the rapid advancement of gene technology over the past 10–15 years, which allows for the transfer of specific genes from a donor genome to the genome of any other species. Anthocyanins (Figure 4.15) or other plant pigments serve as additional identifiers. If only a single isolated protoplast of a distinct origin is fused, markers are not necessary. These two protoplasts are brought together in a small jar using a micromanipulator, and the fusion is frequently induced with an electric shock.

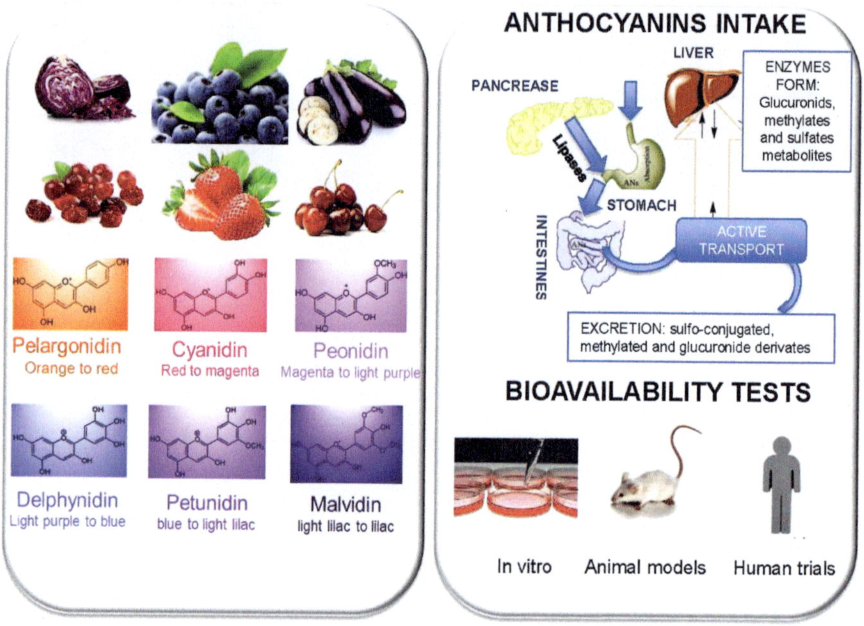

Figure 4.15. Anthocyanins. (Source: Tena N, Martín J, Asuero AG. State of the Art of Anthocyanins: Antioxidant Activity, Sources, Bioavailability, and Therapeutic Effect in Human Health. *Antioxidants*. 2020; 9(5):451. https://doi.org/10.3390/antiox9050451, Creative Commons License)

4.8 PROTOPLAST

The protoplast, which is generated by the plasmalemma, has an overabundance of negative charges on its surface. This prevents two protoplasts from naturally attaching to one another. After over 40 years of protoplast culture research and several attempts to solve this issue, only two approaches are usually regarded as actually workable(Davey et al., 2005). In one instance, significant quantities of macromolecular polyethylene glycol (28%) are used, whereas electroshock is used in the other. Both procedures can be used in different ways and at different costs; for the polyethylene glycol approach, a standard laboratory will do, but for electrofusion, the original home-made equipment has been replaced by commercially produced equipment. The original plant material grew at a specific temperature, and fusion experiments performed better at lower (10 °C) than higher (25 °C) temperatures. This seems to be connected to the fluidity of membranes, which is dependent on their composition, especially for membrane lipids. The goal of protoplast fusion between two plant species is to create new genomes that contain the genetic material from both parental species. Applying X-rays to one parent in this case is one approach to make a new genome. As a result, chromosomes are damaged and partially destroyed in subsequent cell divisions. Haplotype material is frequently used. It is currently possible to implant specific genetic material, such as single genes, using gene technology. In addition to combining the genetic data from the nuclei of protoplasts from various sources, plastids and mitochondria can also be combined. The mitochondria appear to extend before initially dividing, leading to an increase in population.

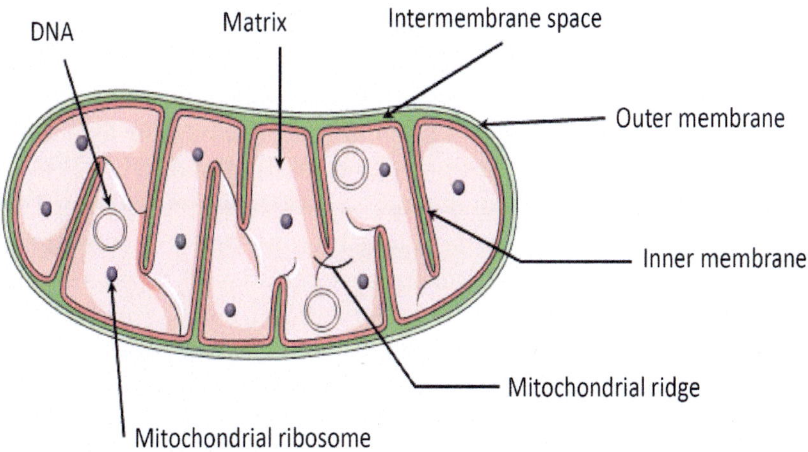

Figure 4.16. Mitochondria. (Source: Andrieux P, Chevillard C, Cunha-Neto E, Nunes JPS. Mitochondria as a Cellular Hub in Infection and Inflammation. *International Journal of Molecular Sciences*. 2021; 22(21):11338. https://doi.org/10.3390/ijms222111338, Creative Commons License)

The mitochondria (Figure 4.16) then disperse uniformly throughout the cytoplasm as a result of an actin-dependent process. This was seen in the first 4–8 hours of protoplast culture following fusion; after 24 hours, the mitochondria of the fused protoplasts had completely mixed. To daughter cells, the mixed mitochondria population is transmitted. It was observed that there will be the formation of haploid plants from anthers harboring immature microspores while conducting a thorough assessment of reactions of different parts of flowers of *Datura innoxia* cultivated *in vitro*. Many research organizations conducted in-depth studies on tobacco anthers, mostly using microspores of this species to test the potential for hybrid breeding projects. When the microspores and their resultant embryos are transformed, homozygous lines (double haploids) are produced that may be used for subsequent breeding. The presence of homozygous parent lines is a need for using the heterosis effect in hybrid breeding. Such inbred lines require numerous backcrosses of heterozygotic parent material to be produced. For plant species that outbreed one another, inbred lines with the desirable traits are also necessary. Haploids allow for a significant reduction in the amount of time needed to create such plant material. Because haploid higher plants are infertile, a diploidization is necessary before haploids can be used in breeding programs.

A stress factor (or phytohormones) initially causes a reprogramming of cellular metabolism, repressing starch biosynthesis-related gene expression and promoting proteolytic and stress-related gene expression. The ubiquitin 26S proteasomal system, which breaks down molecules, can cause metabolic rearrangement; autophagy is involved in the breakdown and recycling of organelles in both plants and animals via vacuoles and lysosomes. Eventually, the breakdown by-products are metabolized. Ideally, the induction of androgenesis entails isolating microspores from the mother plant and then inducing somatic embryogenesis in those microspores. However, in many instances, the development of a callus, in which embryos develop, can initially be seen. Surface-sterilized flower buds are opened under sterile conditions at the appropriate developmental stage, and the anthers are removed using forceps and deposited onto an agar medium. The first embryos begin to appear two to three weeks later. The first mitosis, which results in the vegetative and generative nuclei, is the ideal stage for tobacco microspores. The vegetative nucleus is where the haploid embryo develops. The growth of the flower bud and the development of the pollen are roughly correlated, offering directions for the best developmental stage to collect the anthers for the experiment.

Results on a few variables that are crucial in influencing the capacity to carry out androgenesis and, consequently, the success of an experimental setting are reported. The model used was tobacco. To begin with, pollen dimorphism in mature anthers was noted. Infertile pollen is embryogenic and may be distinguished from fertile pollen using density gradient centrifugation. The pollen cannot mature in these conditions due to the restriction or disruption of the nutritive tissue of the anther, the tapetum. Application of growth agents that encourage male sterility can also increase the amount of P-pollen, so its increase is not simply attributable to disruption of the anther system. Plants that are genetically sterile in males typically have a high potential for androgenesis. Haploid plants can be produced by cultivating anthers or isolated microspores from anthers under the right conditions. A fertile homozygous diploid plant can be produced from a heterozygous mother plant, depending on the plant type, by treating such plant material with colchicine. Because it significantly cuts down on the time needed to generate inbred lines, this technique is beneficial for hybrid breeding. The activation of recessive genes in haploids to be chosen for plant breeding or gene transfer is another goal of employing microspore cultures.

CHAPTER 5

In vitro Propagation

CONTENTS

5.1 INTRODUCTION

For many years, anthuriums have been grown for ornamental purposes. They have attractive foliage and are rather simple to grow. Anthuriums are grown all over the world, with the United States and the Netherlands producing most of them. The ornamental sector has transformed as a result of tissue culture adoption as a standard method for plant multiplication. The production of disease-free stock material and mass reproduction make this method ideal. Nowadays, almost all pot-type Anthuriums on the market are made using different tissue culture procedures.

Figure 5.1. *Anthurium andreanum*. (Source:https://upload.wikimedia.org/wikipedia/commons/8/8b/AnthuriumAndraenum.jpg, Creative Commons License)

Anthurium andreanum (Figure 5.1) plant regeneration has been accomplished via callus, axillary buds, root explants and direct shoot regeneration from lamina explants. Researchers found that most adventitious Anthurium shoots regenerated in loose clumps in liquid cultures. Over 2,000 different species of begonia have been identified, and they are distributed throughout the tropical/subtropical regions of Central and South America,

Asia, and Africa. Begonias are essential aesthetic plants used as garden plants, pot plants, in hanging baskets, and as plants for indoor décor. They are primarily perennial herbs with thick rhizomes or tubers.

Commercial growers have introduced over 200 species, including *Begonia semperflorens*, *B. tuberhybrida*, *B. elatior*, *B. cheimantha*, and *B. socotrana*. Begonias are simple to grow vegetatively by grafting. The most popular of these traditional multiplication techniques for Begonia are leaf cuttings. Begonia leaves occasionally have a high rate of regeneration. As in the case of *B. hispida Schott. var. cucullifera Irmsch*, unattached leaves may have several leaf-like appendages on their surface. On unattached leaves of *B. rex*, the use of phytohormones, particularly cytokinins, promotes the regeneration of buds. The tobacco thin cell layer (TCL) system, which was first characterized, is a useful model for understanding the mechanisms that regulate morphogenesis and transformation. Numerous dicotyledonous and monocotyledonous species were grown using this technique and it was quite effective. Today, it has emerged as the best method for the quick micropropagation of many plants, especially those that are resistant to *in vitro* regeneration, including some legumes, such as *Psophocarpus tetragonolobus* (Figure 5.2) and *Soja biloxi* (Figure 5.3); and woody dicotyledonous species, such as *Hardwickia binata Roxb*.

Figure 5.2. *Psophocarpus tetragonolobus*. (Source: Bassal H, Merah O, Ali AM, Hijazi A, El Omar F. *Psophocarpus tetragonolobus*: An Underused

Species with Multiple Potential Uses. *Plants*. 2020; 9(12):1730. https://doi.org/10.3390/plants9121730, Creative Commons License)

Figure 5.3. *Soja biloxi*. (Source: Domínio público, https://commons.wikimedia.org/w/index.php?curid=16016, Creative Commons License)

Through TCL culture technology, a large number of shoots or plantlets may be easily obtained in less time, enabling additional investigation into areas like DNA transformation and breeding. Begonia rex petioles were examined utilizing the TCL technique using 6-benzyl adenine (BA) and 1-naphthaleneacetic acid (Teixeira da Silva et al., 2015).

5.2 ORGANOGENESIS

Organogenesis was used to obtain shoot regeneration from petioles on MS medium with NAA and BA. It is essential to boost the effectiveness of shoot regeneration, as seen in Begonia, to produce dependable, workable gene transfer and propagation systems. The process of micropropagation depends on some variables, all of which should be monitored. This procedure, which begins with petiole TCLs, has been demonstrated to be effective for regenerating begonia plantlets and can be used successfully in future genetic transformation investigations. This technique can also be standardized for mass propagation of other species and subsequently applied to other economically significant flower species if high-quality shoots are readily available. The orchid species in the genus Vanda are grown extensively as cut flowers in Thailand and Singapore, as well as throughout tropical South and Southeast Asia, New Guinea, and North Eastern Australia. Using a few species in a hybridization effort, both within Vanda and in combination with allied genera, a sizable export industry has emerged.

Figure 5.4. *Vanda coerulea*. (Source:https://upload.wikimedia.org/wikipedia/commons/b/b4/Vanda_coerulea_Orchi_6054.jpg, Creative Commons License)

At the University of Singapore, the first Vanda tissue culture studies were performed using callus made from seedlings in a media containing

2,4-Dichlorophenoxyacetic acid (2,4-D) and tomato juice. Regarding explants, medium composition, and ambient conditions, several Vanda micro cloning techniques show significant results. To encourage the development of shoot primordia, researchers cultured the shoot apices of *Vanda pumila* in Gamborg B5 liquid media supplemented with naphthaleneacetic acid (NAA) and 6-benzyl amino purine (BA) at varied doses. After culturing of shoot primordia in B5 liquid medium supplemented with 0.02 mg/L BA, plantlets were grown 120 days later. Through the culture of mature plant shoot tips and leaf bases on medium supplemented with 10% CW and 0.5-1.0 g/L casein hydrolysate, researchers devised a dependable method for the rapid multiplication of *Vanda coerulea* (CH) (Figure 5.4). On media enriched with 35 g/L banana pulp, 30 percent CW, and 1.08 mM NAA, explants with proliferating buds/PLBs were subcultured.

Figure 5.5. *Ranunculus asiaticus*. (Source:https://upload.wikimedia.org/wikipedia/commons/2/23/Persian_Buttercup_01.jpg, Creative Commons License)

After 30 weeks of culture, the maximum number of shoots were grown from a single shoot/leaf explant. In a medium containing 1.08 mM NAA and 30 g/L banana pulp, rooting was accomplished in 3–4 weeks. A micropropagation system was produced using foliar explants of *Vanda coerulea* grown on VW medium. It was effective to apply a combined NAA and BA therapy to encourage PLB-mediated regeneration in 100% explants. Kinetin (Kn) encouraged the direct growth of shoot buds in place of BA. Due to its great commercial value, vanda teres are now in danger of going extinct. This justifies the production of an effective technique for mass clonal replication, as well as the subsequent commercial exploitation and protection of this significant native orchid of tropical Asia.

The decorative plant *Ranunculus asiaticus L.*, (Figure 5.5) which is a member of Ranunculaceae is a popular family of potted plants and cut flowers. It is indigenous to the eastern Mediterranean region. Turkestan, from where it underwent migration to the western portion of Japan, Israel, South Africa, California, and Europe. The breeding efforts are highly complicated, and population homogeneity cannot be assured. Plantlets develop from seeds, and produce tuberous roots at the end of the growing season; these can be employed in the second year (winter). It is possible to multiply by dividing the tuberous roots, although it does not guarantee disease-free plants, and requires moderate rates of multiplication(Sugiyama, 1999).

5.3 MICROPROPAGATION

It has been demonstrated that micropropagation promotes subsequent dissemination of breeding projects to spread the paternal lines further for the speedy commercial production of fresh cultivars. Growers can be given access to genotypes that are more profitable. There is a need for research to determine the suitability of somatic embryogenesis as a substitute method for mass propagation. Medium-sized deciduous tree *Crataeva adansonii D.C. Prodr.*, of family Capparaceae, sometimes known as Barna or Narnohi is popular due to its beautiful flowers, it is frequently planted as a tree along roadsides and in gardens. The flowering season begins in late March. In addition to ornamental appeal, it also has medicinal and culinary value. The leaves of this plant are used as a tonic, rheumatism remedy, and for stomach aches. Its bark is used to treat fever and nausea. Natural reproduction happens through seeds and roots. After fruits break, seeds covered in pulp are attacked by insects and fall to the ground, further complicating seed dissemination. Explants from mature trees have difficulty being reproduced using tissue culture techniques. These tissues typically respond better to *in vitro* treatment than mature tissues. In addition to the age of the tree, the genotype, the physiological state of the tissue, and the season in which the explants are harvested and cultured are the main factors that affect how the explants respond. Organogenesis, embryogenesis, and axillary proliferation are the three main techniques for *in vitro* plant regeneration that has been used. The advantage of micropropagation via axillary and adventitious branch proliferation is that there is no callus phase in between, preserving the genetic stability of the offspring plants. However, de novo shoot production through callus culture is also preferred for increasing quantity because callus-induced plants are more.

Figure 5.6. *Pitcairnia feliciana*. (Source:https://upload.wikimedia.org/wikipedia/commons/thumb/7/7a/Pitcairnia_heydlauffii_HabitusLeavesInflorescences_BotGardBln0806a.jpg/1200px-Pitcairnia_heydlauffii_HabitusLeavesInflorescences_BotGardBln0806a.jpg, Creative Commons License)

The Bromeliaceae family consists of 56 genera and about 3,146 species and subspecies; more than half of them are epiphytes, meaning that they get their moisture from the air rather than from the ground. Except *Pitcairnia feliciana*, (Figure 5.6) which is native to Africa, they are all native to the Americas. High levels of genetic variation can be found in the Brazilian Atlantic Forest biome. This biome depends heavily on bromeliads, and the South and Southwest of Brazil typically host species of Vriesea among them. There are 257 species of Vriesea spread across the American continent. 31 native Vriesea species, some of which are endemic, have been observed in Santa Catarina State, South Brazil. The development of several bromeliad species in organogenesis has been linked to explant sources in Vriesea. These formations, which are frequently separated from one explant, may respond to stimulation and produce shoots from nodular cultures. Techniques for tissue culturing are useful resources for mass bromeliad reproduction. It's been demonstrated that the Bromeliad leaf basal portion displays vascular elements whose activation by some agents can make cells capable of re-differentiation. Such explants were utilized with success in many *in vitro* bromeliad systems. Clusters of nodules, typically coming from the basal region were induced. The ability to micropropagate through tissue culture allows the regeneration of several plants from fragments (explants) of stock plants in a shorter time and in all seasons(Filho et al., 2005).

The introduction of a successful *Agrobacterium*-mediated transformation technique will enable the flower to acquire the novel trait. There have been reports on several genetic modification procedures for flowers. However, the lack of an effective procedure restricts the genetic modification of flowers. Low transformation efficiency is one of the serious issues with floral genetic transformation. The strain of *Agrobacterium* utilized, the health of the plant, inoculation techniques, and co-cultivation techniques are only a few of the variables that can affect the outcome of genetic transformation. Explants of flowers from greenhouse plants were more vulnerable to *Agrobacterium* than explants from *in vitro* cultures. Numerous studies suggest that phytohormone explant preculture is essential for effective genetic transformation. The most crucial phase in the genetic modification of plants is cocultivation. Some researchers have successfully transferred genes to explants of plants like *Cyclamen persicum* and tomato that were co-cultivated with a hormone-rich media. For example, for *Drosera rotundifolia L.*, some have reported successful genetic transformation by lowering the content of the co-cultivation medium to half that of the regeneration media. Studies have shown that acetosyringone (AS) is a crucial compound that influences the success of genetic transformation. Pre-treatment of explants with AS in the S phase of the cell cycle could accelerate bacterial adhesion to cell walls. AS activates the virulence genes of *Agrobacterium* and enhances the transfer of foreign genes into plant genomes. The way by which *Agrobacterium* attaches to plant cells and how this impacts the success of the genetic transformation is of great interest to researchers. The resulting large colonies of bacteria are subsequently surrounded by a cellulose fibrillar net. The bacteria become bound to the surface of the plant cells as a result of these fibrils. The ability of *Agrobacterium* to alter its genetic material can occasionally be affected by the attachment of bacteria. The transmission of T-DNA requires the adhesion of bacteria to plant cells, and research indicates that these attachments are polar.

To improve flower transformation efficiency, explant preparation and co-cultivation medium are both crucial. Due to two essential factors, the liquid culture approach has enormous potential to enhance shoot organogenesis. One is that a liquid medium enhances the oxygen diffusion coefficient, and the other is that a liquid medium maintains a higher level of water potential. The aeration levels and continuous flow of nutrients and growth regulators to the tissues are maintained by these two key characteristics. However, there is a risk of apoplastic deposition of water inside the cells when the cells and tissues come into direct and prolonged contact with the liquid media. Hyperhydric tissues are those that have been impacted in this way. The hyper hydric plants are resistant to *ex vitro* transplanting and are prone to wilting. By placing an inert absorbent or adsorbent support material between the tissue and liquid medium, it is possible to reduce the occurrence of hyperhydricity while maintaining a continuous medium supply. The polypropylene membrane was one of the suggested support matrices that proved particularly helpful in boosting shoot proliferation and lowering hyperhydricity in a variety of monocotyledonous ornamental plants. Despite controlled conditions, cultural variations are unavoidable. In closed vessels, CO_2 and humidity gradients occur because there is little air movement. Additionally, light intensity fluctuates, especially in thin vessels. The growth and uniformity of the quality of the regenerated plants are both impacted by these irregularities. Hands-on identification of the inherent and nonlinear fluctuations in plant cell and tissue cultures is challenging. A neural network-based computer vision system can provide a useful solution for categorizing these groups. The ability to organize plants into homogenous groups allows us to choose species or groups of plants based on important biological characteristics.

Figure 5.7. *Dianthus caryophyllus*. (Source:https://upload.wikimedia.org/wikipedia/commons/3/3b/W_carnation4051.jpg, Creative Commons License)

Dianthus caryophyllus L., (Figure 5.7) a member of the Caryophyllaceae family, is the most popular ornamental plant in the world and a prime candidate for breeding new varieties with enhanced agronomic properties. The floriculture industry values their excellent aesthetic value. Due to the large percentage of endemic species in this genus, many of which are listed on national and European Red Lists, it has significant phytogeographic significance. Additionally, *D. caryophyllus* is frequently used in physiological research, particularly about the senescence of flowers. Therefore, the availability of a practical and simple method of flowering plant propagation via *in vitro* techniques may be of significant relevance for a variety of uses. Hyperhydricity usually hinders *in vitro* flowering culture. Atypically, the leaves or shoots develop a water-soaked appearance. The phenomenon, which was recently renamed as hyperhydricity, was previously known as vitrification or glassiness. To meet the rising demand from industry, ornamental plants, notably aromatic species, which play an important role in world trade, need to produce at a faster rate. Commercial growers accept tissue culture plants in addition to resources from the wild. All perfume enthusiasts know about jasmine flowers; their authentic scent cannot be replicated. The premium species utilized in the industry, *Jasminum officinale*, (Figure 5.8) is a woody vine that grows to a height of 10-15 feet when fully grown and requires support to ascend by twining(Jack et al., 2005).

Figure 5.8. *Jasminum officinale*. (Source:https://upload.wikimedia.org/wikipedia/commons/thumb/b/b9/Jasminum_officinale_jd_Bout.png/800px-Jasminum_officinale_jd_Bout.png, Creative Commons License)

One of the focus areas for flower crops in India that has drawn more attention from the government is the highly commercial *Jasminum officinale*. The focus has mostly been on crop development and agro-technique standardization, including mass propagation for planting materials needed for cultivation in wider regions. A barrier to meeting the demand of the fragrance industry is the lack of sufficient planting material for extensive cultivation. Although tissue culture techniques are in place for a while, nothing has been done to improve replication rates in woody, aromatic, and decorative plants. Examples of attempts at micropropagating woody ornamentals include callus culture, protoplast culture, organogenesis, and somatic embryogenesis. Woody plants can be multiplied quickly by promoting adventitious bud growth and axillary branch proliferation, which is facilitated by the application of the correct phytohormones. The most reliable method to maintain the genetic integrity of a clonal population is to stimulate axillary branching; however, due to the heterogeneity of seeds, this tissue culture-based propagation strategy may not maintain the genetic integrity of the population. There are over 2,000 species in the family

Gesneriaceae, most of which are grown in tropical and subtropical regions of the world. The family contains a variety of plants with vibrant flowers and high ornamental value. Many wild species and cultivars from some genera, including Achimenes, Aeschynanthus, Chirita, Gloxinia, Saintpaulia, and Streptocarpus, have been cultivated. Due to habitat loss and excessive collection for horticultural purposes, many wild species have become endangered. Recently, it has been acknowledged that protecting endangered species and using them sustainably are global tasks.

Therefore, it is critically necessary to build breeding and propagation methods for Gesneriaceae species that are kept in botanical institutes or gardens as genetic resources. Numerous studies have already documented the micropropagation of plants in the family Gesneriaceae, particularly those in the genus Saintpaulia. There are eight subgenera in the genus Rhododendron, which contains about 1,000 species. The four most significant subgenera are Tsutsusi, which includes 117 species of evergreen azaleas, Pentanthera, which includes 30 species of deciduous azaleas, Rhododendron, which includes 542 species of lepidote rhododendrons, and Hymenanthes, which includes 302 species. Due to their ornamental significance, these subgenera are significant economically: the latter two are what are colloquially referred to as "garden rhododendrons," whereas Tsutsusi and Pentanthera species are the parents of several hybrid azalea groups that grow both indoors and outdoors. The subgenus Rhododendron contains 310 species of tropical Rhododendron. Most Rhododendron species are typically propagated using stem cuttings. However, after the invention of shoot-tip culture, *in vitro* micropropagation steadily gained popularity and is now playing a bigger role in commercial production.

Efforts have been made to create effective protocols. There are described protocols for Rhododendron micropropagation, organogenesis, somaclonal variation, polyploidization, and interspecific hybridization. It is challenging to develop *Anthurium andreanum* regeneration systems utilizing explants obtained from plants kept in greenhouse environments. To prevent genetic heterogeneity between the regenerated plants, micro cuttings can be taken from these plants and the micropropagation process can be constructed. Although *A. andreanum* seeds can germinate in the dark, we can observe that after one week, 74 percent of seeds germinated under constant light, which is a significant improvement over the 30 percent of seeds that germinated in the dark on MS medium with 2 mg/L BA. One week later, the radicle appeared, and the shoot expanded in the presence of constant light while failing to do so in the absence of light. To provide an adequate number of replications for statistical analysis, we advise cultivating at least 50 micro cutting or callus fragments. To understand the growth of meristemoids, shoot primordia, or roots from these callus tissues, it is critical to examine the organogenic process from callus tissue at the microscopic level. Only callus fragments with a diameter no greater than 1 cm should be used. To have the proper orientation to cut the callus fragments, several safranine pellets might be added. Because the alcohol may entirely evaporate on the first day during paraplast inclusion, it is preferable to cease agitation and add extra paraplast before leaving the mixture overnight. This will prevent the loss of samples.

5.4 SAMPLE PREPARATION

The samples must be suitably oriented, and bubble formation must be prevented. Xylol needs to be used frequently to clean the blade. The 50/50 mixture of glycerol and egg whites must be utilized immediately and filtered through filters. To prevent contamination, some phenol crystals must be added to this solution. To make slide transfer during the stain treatment simple, a slide jar with a glass cover should be employed. The rate of shoot regeneration is also greatly influenced by TCL thickness. TCL should be three millimeters thick if you want uniform shoots and a high frequency of shoot formation. If petiole TCLs are grown in media with only BA, the caulogenesis rate may be lower than in media with BA and NAA, and the shoot formation rate may be lower than 60%. Caulogenesis is observed eight weeks later. Caulogenesis from Begonia petiole TCLs is effective at low concentrations of TDZ. When using 1.0 mg/L BA, the maximum shoot formation was observed. Shoot development is greater than 75% when BA and NAA are employed for petiole TCLs at low doses. VW macro and micro salt stock solutions can be kept at 10° C for up to four months. To ensure the solution is fully thawed before taking any aliquots, stock solutions of auxins, cytokinins, and vitamins are kept at 20° C. To eliminate germs and fungi, the potting mix, which is made of coconut husk and charcoal chips, should be autoclaved before use. Because roots in Vanda trees grow from every node, plants that are fastened to logs of wood so that the roots can adhere to them do very well. Because *Ranunculus asiaticus* can contract a variety of viral diseases, the production of disease-free plants is a serious concern. An integrated production plan should be carried out to ensure virus-free starting material for healthy large-scale production as there are currently no effective ways to remove viruses by the meristem culture of

Ranunculus(Yeung, 1999). Micropropagated plants exhibit noticeable vegetative development in the field. They begin three months after planting than that of seedlings of the same age, albeit the degree of development is genotype-dependent. *Ex vitro* plantlets typically exhibit homogenous flowering, and more than one tuberous root at the end of the culture cycle, albeit this production in a plant is genotype-dependent. The hardening phase should be carried out in Mediterranean conditions from October through January to ensure effective tuberous root production. The acclimation time may affect both the size and production of tuberous roots. In general, the propagation approach using axillary bud stimulation results in production. Nevertheless, suitable selection is still advised for each micro propagated line. Continuous subculturing must be taken into consideration because if *in vitro* culture is carried out for more than 12 to 22 months, a lower flowering percentage and the emergence of some abnormalities can be noticed in some lines. The cost of *in vitro*-produced plants is higher than that of traditional propagation units and tuberous roots from seedlings; however, many advantages can be anticipated: micropropagation aids in providing growers with better performers and healthy genotypes, and an improved production schedule can be anticipated for cultivation and packaging.

MS vitamin supplies and trace amounts of salts are kept frozen; therefore, all solutions must be fully thawed before use. At 4° C, you can store any prepared stock solutions of vitamins. Each flask may contain no more than five explants. On the rooting media, 90 to 100 percent of the regenerated shoots can take root. You can use NAA (1 mg/L) to help the regenerated shoots take root. On NAA-containing mediums, root initiation typically occurs in two weeks. Plantlets that are prepared for transplanting into pots have a minimum of 6-7 roots that are 3–4 cm long and have 4-5 nodes each. We need to remove any traces of the agar medium from the plantlets. We need to use new or recently cleaned and disinfected pots. To eradicate the fungi that are present in the soil, add fungicide to the potting mix. Transfer the plantlets one at a time into pots containing a 1:1 ratio of garden soil to farmyard compost, and then let them harden in a room so that light can enter and a high humidity level can be maintained. Through intermittent exposure to the low relative humidity of *ex vitro* settings, the plantlets are gradually acclimated.

Vriesea friburgensis (Figure 5.9) grows predominantly in sandy and dry soils found on the seacoast, where the general vegetation is typically that found on dunes, and are found in epiphytic or terrestrial habitats in the southern and southeastern Brazilian parts of the Atlantic Forest domain.

Although they are primarily halophytic, they can survive in diffuse light and have few humidity requirements.

Figure 5.9. *Vriesea friburgensis*. (Source: Lima, José & Soares-Silva, Lucia. (2016). Two new records for Bromeliaceae in the central-west region of Brazil: *Vriesea friburgensis* and *Tillandsia polystachia*. Rodriguésia. 67. 1093-1100. 10.1590/2175-7860201667419., Creative Commons License)

This bromeliad faces extinction in its native habitat. Therefore, it is essential to create plans for its preservation and widespread propagation. The endemic bromeliad *Vriesea reitzii* from South Brazil is an ornamental plant. This species has inflorescences that are red and yellow in color. It grows in areas between 750 and 1,200 meters above sea level in the Brazilian states of Santa Catarina, Paraná, and Rio Grande do Sul. *Vriesea splendens* has linear, green, or purplish foliage that is 30–45 cm long and 2.5–4 cm broad. Smooth edges and perhaps colorful bracts can be found at the bases of the leaves. The leaf frequently forms an arching rosette that resembles a funnel. An upright inflorescence with short-stalked yellow flowers that are arranged in two rows of flattened spike-like racemes or panicles and bright red or orange overlapping bracts arises from the center of this rosette. While the bracts are colorful, the petals are frequently tubular and short-lived. It is a terrestrial plant that grows as an epiphyte in tropical Trinidad, Venezuela, Guyana, and Surinam(Corredor-Prado et al., 2015).

The huge bromeliad species *Vriesea fosteriana* is found from Mexico to Brazil. It displays large, mid-green leaves with a reddish-brown band. This leaf can reach a length of 1 m, forming a dense rosette from which a prominent flower stalk rises to 1.5 m, producing a yellow flower from each bract. *Vriesea hieroglyphica* with exceptionally attractive foliage, which features glossy, brilliant green leaves with broad, wavy horizontal bands of dark green, is what gives it most of its ornamental significance. The leaves measure 7.5 cm wide and 90 cm long. One of the world's biggest bromeliads, the imperial bromeliad (*Alcantarea imperialis syn. Vriesea imperialis*) grows naturally in the rocky southeast of the Mata Atlantica. The height of this amazing, magnificent big bromeliad, which grows to a maximum of 2 meters, is between 90 and 120 cm. The massive inflorescence bears white flowers. This bromeliad may take 50 years to bloom in its native habitat, but in cultivation, it only takes 5-7 years. One of the most widely grown bromeliads in gardens nowadays is *A. imperialis*. *Dyckia distachya* has several ornamental characteristics. It can be found in the Brazilian Atlantic Forest, which is a hotspot for biodiversity and one of the most significant centers for plant diversity. In this biome, bromeliads are common, with the majority of them being indigenous. Ornamental plant reproduction has been done using temporary immersion systems (TIS). Their advantages in organogenic regeneration systems are linked to higher multiplication rates (shoot proliferation) and better acclimation responses. These methods lower production costs by minimizing manipulation, labor, space, and the quantity of containers.

The majority of Euphorbiaceae species are distinguished by the production of milky latex, which is frequently poisonous and irritates the skin. When handling poinsettia plant material, it is advised to wear disposable gloves to prevent skin irritation. Stock solutions of cytokinins and auxins, each at a concentration of 1.0 mg/mL, should be made separately. The compounds should be dissolved in tiny amounts with either 1.0 M KOH or 1.0 M HCl. Utilizing reverse osmosis water, adjust the volume. Growth regulator solutions should be kept at 4° C in the dark. Every two weeks, stock solutions should be changed. Stem cuttings are obtained from stock plants of poinsettias. Take the cuttings, which should each be 5-8 cm long and have 2-3 mature leaves, and place them in compost in plastic pots with a 9 cm diameter. Water every three to four days and maintain glasshouse conditions until explants are needed. Axenic cultures of poinsettias may be challenging to generate since the source material is frequently contaminated with microbes. Surface sterilization with a 10 percent bleach solution after immersion in 70% ethanol often removes contaminated bacteria. Explants taken from donor potted plants older than two months old are more likely to produce large amounts of white latex. Significantly, this latex has an impact on the shoot regeneration of cultured explants. Use extra caution while heating culture medium-containing bottles in a microwave. The lids should be loosened.

Figure 5.10. Poinsettia. (Source: Guo Y, Niu G, Starman T, Volder A, Gu M. Poinsettia Growth and Development Response to Container Root Substrate with Biochar. *Horticulturae*. 2018; 4(1):1. https://doi.org/10.3390/horticulturae4010001, Creative Commons License)

5.5 SHOOT REGENERATION

The poinsettia shoot regeneration (Figure 5.10) from stem nodal explants is genotype-specific. Explants from donor plants of other colors, such as cultivars with white bracts, typically produce fewer shoots than explants from cultivars with red bracts. However, all of the poinsettia cultivars were successfully able to induce shoot regeneration from at least 70% of the explants. Some regenerated buds might not grow, in which case these cultures ought to be thrown away. Compared to Petri dishes, the headspace above the culture medium is higher, promoting the growth of regenerated shoots. The recommended auxin for producing roots on poinsettia regenerated shoots is indole-3-acetic acid (IAA). It is advised that longer roots be cut to 2-3 cm in length and that any dead or damaged lateral roots be removed before transferring regenerated plants to compost. In general, poinsettia plants can be kept alive *ex vivo* in a variety of soil replacements, such as perlite, vermiculite, and sand with the addition of organic material like peat, humus, and bark. Poinsettias are categorized as short-day plants, and under ideal glasshouse conditions, they grow colored bracts after 8 to 9 weeks with a photoperiod of 12 to 12.25 hours. MS macro and micro-alt stock solutions can be kept at 10°C for up to four months at half strength. Stock solutions for auxins, cytokinins, and vitamins are kept at 20°C, which guarantees that the solution has fully defrosted before any aliquots are taken. Before using the potting mix (coconut husk and charcoal chips), bacteria and fungi should be killed in an autoclave(Huang & Chu, 2008).

Plants should be kept in a shaded, well-ventilated house. The most vital thing to keep in mind is that these leaves may quickly burn if they are exposed to excessive sunshine. When preparing node explants, it's crucial to tear rather than cut away leaves from the shoot since adventitious shoots will only grow from the scar left by the ripping, not the cutting. When carefully separating leaves from shoots to prepare leaf explants, make sure that the explants contain the connecting line between the leaf and the shoot since adventitious shoots regenerate from this line. For leaf explants, explant preparation was a crucial step in obtaining the plants that had undergone a transformation. The cut ends of the leaf explants were the only parts that stained with GUS. Only explants with a cut end at the leaf base give rise to GUS-positive shoots. Node explants from greenhouse plants produced a higher number of transformed shoots than leaf explants. The size of the explants and their capacity for regeneration may be the cause of the differential in transformation efficiency between the node and leaf explants. Greenhouse node explants generate ten times as many adventitious shoots as *in vitro* leaf explants. As a result, there is a greater likelihood that the node explants will result in altered shoots. The *in vitro* materials offer certain advantages despite having a relatively poor transformation efficiency because transformants from *in vitro* starting materials are anticipated to be clean. Clean materials were needed for several genetic transformation investigations, such as an attempt to insert virus-resistant genes into plants.

To eliminate the background effect during the staining procedure, nontransformed explants should always be used as negative controls when doing GUS experiments. This procedure is repeatable and adaptable to various flowering genotypes. When applied to transferring genes into *Dianthus hybrida* regenerable calli, this approach was still effective. All stock solutions for media preparation are kept at a temperature of 4°C. Before use, Duroplast foam is properly rinsed in double-distilled water, steeped in a warm solution of 0.1 percent (v/v) liquid detergent for at least 2 hours, and then dried in an oven. The training set ought to contain a lot more leaf patterns than the test set. The number of patterns in the validation set must be comparable to the number in the training set. When using apical buds, choose shoot tips from the apex that are 2 cm long. To prepare explants, stems are cut around 0.6 mm below the apical dome. In place of agar, a solution of 0.4 percent BDH and 0.4 percent Roth agar can be employed. You can also employ incident radiant flux up to 50 mmol/m^2/s. To accelerate shoot multiplication, agar strength may occasionally be increased by up to 1.2%. Internodal callus tissue could be employed. Losses of up to 60% have been documented in various commercial micropropagation systems due to hyperhydricity, a complex reaction and major problem. Although numerous findings have indicated the existence of oxidative stress, deficits in the production of cuticular waxes, or alterations in the methylation pattern of pectins in cell walls, the physiological basis of this alteration has not been well defined.

The needs for metal ions vary across different flowering types. For each situation, the metal supply needs to be optimized. By bottom cooling system, we need to cool the base of vessels, allowing the inner atmosphere of the vessel to condense into the medium. This is simply accomplished by circulating water that has previously been cooled by a cryostat together with anti gel fluid. Containers are put on metal shelves. The relative humidity within the culture flask is 75% when the temperature of the circulator is set to 20°C and the ambient temperature is 25°C. Pantomycine is soluble in water while bavistin is not. To make a solution with these two antimicrobial substances, weigh each substance separately based on the concentration, stir in 25–30 ml of distilled water for 5 minutes, and then add the remaining amount. We can get a turbid solution. Use of 1 N NaOH or 1 N HCl to modify the pH of the medium before adding agar to it is advised. Before altering a new sample medium, calibrate the pH meter each time. Glassware and culture media should be autoclaved for 20 min at 121°C and 103 kPa. To make stock solutions of BAP and kinetin, dissolve the crystals by adding a few drops of 0.1 N HCL, then add the necessary amount of liquid to a volumetric flask. In the same manner, dissolve crystals in 0.1 N NaOH to prepare IBA and NAA. To make MS and Miller's medium at half its normal concentration, merely reduce the inorganic salts by half, leaving the remaining ingredients unaltered. Prepare a filter paper bridge by folding filter paper into a 5-layered, 1 cm wide strip. Then, gently position the strip at the base of the culture tube containing the liquid rooting media. At various locations, perforated poly bags and cover the plantlets-containing plastic cups. The poly bags shield young plants from the outside air, reducing transpiration rates and allowing atmospheric gasses to enter through tiny pores. Specifically, March to April is the best period to harvest explants since *Jasminum officinale* begins to grow vigorously at this time after going through a lag phase in the winter. After this time, plants experience significant hormonal changes, begin to flower when the majority of the vegetative buds transform into flowering buds and are consequently unsuitable for forced branching and *in vitro* culture. Spraying with antimicrobial agents decreases the high risk of deteriorating explants following inoculation, making it a required step before *in vitro* use. Avoid choosing fragile young branches since the cut ends of explants from these portions begin to form calli early in the culture process. The intensity of brown leachate reveals the subculture formation. It normally takes two successive cultures to halt leaching, but if more adjustments are required, they should be made. In a shoot multiplication experiment, use separate culture tubes for each explant since a single culture suffers less loss from contamination than several cultures in flasks do. From a culture tube to a bigger container, good cultures can be transferred.

To ensure that growing roots may penetrate the base of freshly developed regeneration shoots without difficulty, leave at least a tiny amount of hard tissue there. When cleaning and removing the agar from the roots, take extra care to preserve the root tips. When transferring to a solid substrate, place the roots appropriately so that they do not attach. China-native *Lysionotus pauciflorus* is a diploid with 32 chromosomes. For the germination of these seeds, the ideal temperature is 25°C, and light is essential. As previously observed, single-node cultured plants were genetically stable. Leaf explants were cultivated on media supplemented with NAA and 2,4-D, each at a concentration of 1 M, and the result was the induction of embryo-like structures. Despite their initial appearance as somatic embryos, histological examinations revealed that these structures were monopolar, and upon development on the same medium, they transformed into small roots with root hairs. Therefore, these entities were root primordia rather than somatic embryos. 2,4-D at high concentrations prevented leaf tissue callus and shoot induction. The cell culture system may also be used for somatic hybridization and genetic transformation investigations because it has maintained a high capacity for plant regeneration for more than two years. Protoplast culture and genetic transformation are known to be difficult in Gesneriaceae plants, but both can be accomplished by using callus culture, which has a high capacity for plant regeneration. Regenerated shoots are readily rooted without the use of plant growth regulators on a half MS media(Ebert & Taylor, 1990).

Although shoots with a few variegated leaves frequently regenerated from callus, they did not continually produce variegated leaves during follow-up *in vitro* and *ex-vitro* cultivation. A flow cytometric examination revealed that 8 out of 66 acclimated plantlets were tetraploid, the majority of regenerants from callus were diploid. Due to changes in morphological and physiological traits like longer shoots, thicker stems, deeper green leaf color, larger and thicker leaves, and larger flowers, tetraploids are occasionally helpful in ornamental plants. They could develop into the best source of ornamental plants with significant added value for the floriculture sector. A sequential treatment using 30 percent H_2O_2 (15 min), 7 percent NaOCl (10 min), and 0.1 percent $HgCl_2$ (10 min) with 0.1 percent Tween 20 added to each solution is an alternative method for significantly contaminated material. Before NaOCl therapy, ultrasonication might be used for one minute. Floral structures, seedling cultures, and *in vitro* meristems can also be used to initiate multiplication. Cytokinin that is used the most is 2iP. With yielding rates of 7–20, it is more effective than BAP and less expensive. Although TDZ substantially encourages shoot proliferation, the resulting shoots are frequently deformed, compressed, hyperhydrated, and unsuitable *in vivo*; as a result, it is typically employed for regeneration rather than multiplication. IAA or NAA are occasionally employed to counteract the effects of cytokinin and to produce more strong shoots.

5.6 PRETREATMENT

Although there are various callus varieties, hard, compact callus is typically preferred to soft, friable callus that rarely regenerates. Various cytokinins can be applied. Because flowers have a broad margin that has a distinct color from the rest of the petals and is made of tetraploid tissue rather than diploid remnants, only these cultivars are appropriate. Rhododendron genotypes from the same subgenus can hybridize with other species of the same genus. Research and progress in many areas of plant science have improved because of the use of micropropagation, which has made it possible to undertake fundamental studies on the physiology of morphogenesis and differentiation. Tissue culture has made it possible to mass-produce superior genotypes of ornamental plants, enabling the commercialization of healthy and consistent planting material. Furthermore, *in vitro* culture opens up new possibilities for plant enhancement when combined with sophisticated technologies. Passiflora contains over 500 species of vines, lianas, and tiny trees. The genus is found across the warm temperate and tropical parts of the World, with a few Malayan, Chinese, North American, and Australian species being an exception. For their edible fruits, several species are grown in the tropics, while many others are grown either outdoors, in warm climates, or in glasshouses. The floriculture industry is particularly interested in Passiflora species because they have various distinctive floral characteristics, such as multiple series of vividly colored coronal filaments and intricate floral nectary structures(Alleweldt & Radler, 1962).

Passiflora species and hybrids are sold as garden pot plants in the Italian floricultural market after being reproduced using *in vitro* culture methods. The Passiflora hybridcalled 'Guglielmo Betto' is important due to its intriguing ornamental characteristics, including its large, double series of colored filaments, quick growth and profusion of flowers, and resistance to low temperatures. Systems for effective shoot multiplication and regeneration were devised. Early research on the physiological effects of submerging the tendril tip in a solution containing various auxins on *Cucurbita moschata* and *P. coerulea* tendrils found that IAA, in particular, operates to induce coiling by rapidly engaging with auxin-specific receptors in the tendril tip. Furthermore, two early auxin response genes in pea that are similar to the Arabidopsis auxin efflux carrier have been identified. Regardless of genotype, the PsPK2 mRNA is more prevalent in the tendrils of pea leaves than in the leaflets, suggesting that it may be involved in the thigmotropic response. Tendrils transmit auxin far more effectively than other leaf components. Passiflora tendril tips have high morphogenic behavior due to their effective auxin transport. Different BAP concentrations have been used. The maximum number of new shoots per explant have been produced by adding 0.11 mM BAP. The shoots were ready to be split and placed into elongation and rooting media after two months since they had grown well and extended. Similar results were observed in the number of shoots where the BAP concentration was the highest, the shoots were chlorotic and smaller in size.

One of the most significant ornamental plants in the world is the rose. In addition to being used as a garden plant, it is also cultivated for the production of cut flowers, root stock, and rose oil. It can take on a variety of shapes, including bushes, shrubs, and climbers, and it can be deciduous or evergreen. They also exhibit a wide variety of variations in flower shape, size, color, scent, and other agroclimatic characteristics. In general, vegetative procedures including cutting, layering, budding, and grafting are used to propagate roses. The generation of rootstocks, new cultivars, and species is all accomplished through the usage of seeds. Although roses are typically propagated vegetatively, this method does not always result in strong, disease-free plants. Additionally, seasonal dependence and low multiplication rates are some other significant barriers to conventional propagation. Alternative plant propagation techniques like *in vitro* culture offer special benefits like the ability to produce healthy, disease-free plants and propagules throughout the year. They also multiply quickly in a short amount of time. Researchers have reported micropropagation roses using cultures of axillary buds and apical meristems. There are several steps in the effective micropropagation process for roses, each with its own set of requirements. The steps typically include starting aseptic cultures, multiplying shoots, establishing micro shoots as roots, and hardening and transferring tissue culture-raised plants to the field.

CHAPTER 6

Transgenesis

CONTENTS

6.1 INTRODUCTION

A basic question is where the genes came from that A. rhizogenes (Figure 6.1) was introducing into plants (White et al., 1985). They may come from a variety of sources, including plants themselves. For instance, the genes in the Ri TL-DNA found in wild type plants may have originated as the bacterial counterparts of these genes. Biomass from dead plants reaches the soil environment and is broken down by microorganisms there. Cap cells shed as roots expand, releasing their contents into the rhizosphere. A portion of this freed plant DNA may be ingested by bacteria, integrated into a T-DNA, and then returned to plants since DNA survives in the environment and is imported by bacteria.

Figure 6.1. A. rhizogenes. (Source: https://upload.wikimedia.org/wikipedia/commons/a/a1/Rhizobium_rhizogenes.png, Creative Commons License).

Soil bacteria may operate as a genetic link between plants of different species since T-DNA transfer to plants can take place on a wide variety of hosts (McKnight et al., 1987). Technical challenges were encountered in attempts to show DNA transfer from plants to bacteria, in part because DNA encoding marker genes are found in soil metagenomes and as laboratory pollutants. Acinetobacter baylyi strain of the soil bacterium, Johann de Vries and Wilfried Wackernagel utilized a sensitive approach for measuring the uptake and incorporation of a DNA from an antibiotic resistance gene. The technique was called marker exchange, and it involved replacing a damaged nptII (kanamycin resistance gene) in the target DNA with an intact nptII from the source DNA through homologous recombination. Researchers recorded the transfer of DNA from six plant species into A. baylyi using this technique. Sources comprised complete plants in vitro, intact leaves, and pulverized leaves. The problem of DNA contamination was handled by introducing silent mutations in the source plant nptII, which were detected in the bacterial target. Transfer was dosage dependent and DNase sensitive.

6.2 GENE TRANSFER

It was simple to envision how conjugation, once established in A. baylyi, could transfer a DNA sequence into other bacteria, such as Agrobacterium, which has a DNA transfer system for plants. However, the homology between the entering foreign DNA and the sequences contained by A. baylyi was necessary for DNA integration. The Ri T-DNA already present in the plant host may serve as a source of homology(Chandra, 2012). Therefore, DNA transfer in either direction would offer the potential for transporting additional sequences, such as those interspersed among the plast genes, between plant species via A. rhizogenes. Once the homologous sequences were in place, it was irrelevant to know the initial transfer's orientation. These queries sparked a reconsideration of the rhizosphere's nutrient interactions between plants and microorganisms. Root exudates, root necrosis, and the contents of the cap cells that shed during root growth all provide food for the microbes that dwell around roots. Whether or not the secondary metabolites are encoded in a T-DNA, bacterial catabolism of plant secondary metabolites is adaptive and specific to plant species. A. rhizogenes, which was capable of nodulating legume roots and fixing nitrogen in the nodules, was closely related to Rhizobium. Rhizobia and their legume hosts interact with opine-like compounds.

When testing for agropine in C. sepium transformed by A. rhizogenes using high voltage paper electrophoresis and silver nitrate staining, compounds in the roots of both transformed and control plants were observed(Tepfer, 1983). They accumulated preferentially in the underground organs of both control and transformed plants in quantities comparable to agropine, and they reacted with silver nitrate similarly to agropine and manopine. They might be the only sources of food for plant-associated soil bacteria. The catabolic area on $pRme_{41}$ was characterized as a result of these s investigations. Only three of the 105 plant species that were subjected to screening for calystegins in their underground organs did so: C. sepium, C. arvensis, and Atropa belladonna. The calystegins were purified by A. Goldman, who had found nopaline and octopine, and their structures revealed that they were new tropane derivatives that resembled glycosidase inhibitors. Glycosidases are toxic because they play a crucial role in cell-to-cell recognition, among other things. Additionally, whereas non-producing plant roots lacked the bacteria that catabolized them, the roots of plants producing calystegins did. It seems obvious to understand things ecologically. Only soil bacteria with specific catabolic genes can break down the compounds that plants release as carbon and nitrogen into the soil.

Figure 6.2. Rhizobium. (Source: https://upload.wikimedia.org/wikipedia/commons/2/29/Lotus_pedunculatus11_ies.jpg, Creative Commons License).

Calystegins, however, are also allelopathic glycosidases that destroy organisms and plants in the rhizosphere, making them more than just exclusive carbon and nitrogen suppliers(Défago et al., 2001). The bacterial catabolic genes, which are conveniently carried on a self-transmissible plasmid for propagating among bacteria, would therefore not only offer exclusive sustenance to soil bacteria but also protect the host plant by detoxifying the soil. One aspect of a larger network of nutritional interactions included the opines and T-DNA transfer. The catabolism of calystegin was not required for nitrogen fixation or nodulation. It is likely that R. meliloti (pRme41) was isolated from soil where morning glory had recently grown. Apart from nodulation and nitrogen fixation, rhizobia live a saprophytic life.

6.3 MEDIATORS

Dietary mediators might be closely connected to nitrogen fixation and nodulation. Researchers concentrated on the betaines found in alfalfa seeds because they believed that when the seeds grew, betaines would be released and begin interactions with Rhizobium (Figure 6.2) through nutrient selection. They identified the Rhizobium symbiotic plasmid (pSym), which is surrounded by genes involved in nitrogen fixation, as the site of betaine catabolism. Mutagenesis demonstrated that stachydrine catabolism was necessary for effective nodulation in accordance with the activities of surrounding genes. The catabolic area was divided and sequenced that demonstrated similarities to bacterial genes involved in the detoxification of xenobiotics. As with DNA transfer, they came to the conclusion that dietary interactions are crucial to nodulation, but allelopathy can also play a role. As bacteria and plants coevolved, layers of complexity were added. There was no proof of DNA transfer in the Rhizobium symbiosis, but there was no need for it because a bacterium called A. rhizogenes carried out this function more than adequately. More significantly, the research on nutritional mediators supported the idea that dietary interactions affect the interactions between plants and microorganisms. In the case of opiates, it is evident that the anabolic genes (carrying by the Ri and Ti T-DNAs) are the targets of dietary mediators (such as calystegins and betaines) were also candidates.

Small proteins known as cryptogeins are from Phytophthora, the fungus that caused the late blight that resulted in the Irish potato famine in the middle of the nineteenth century. They secrete a lot of these in order to scavenge sterols. It was once believed that plants used these elicitins as markers for the presence of *Phytophthora cryptogea* and as a cue to activate their defenses, which included producing antimicrobial metabolites. From the purified protein, researchers identified the amino acid sequence of cryptogein. The expression of the gene for cryptogein in a host plant could activate a defence reaction and defend the host, making it a genetic experiment. Some researchers created the DNA sequence that codes for the cryptogein protein. Researchers expressed the resultant gene in yeast after it had been altered to express a valine rather than the crucial amino acid lysine. The wild type -cryptogein from the synthetic gene produced the same necrotic lesions as the pure natural protein when injected into tobacco leaves, however, the mutant induced comparatively minimal necrosis. There was a correlation between varied levels of Phytophthora resistance and the expression of the synthetic cryptogein gene and the mutant in tobacco when they were both expressed under a constitutive promoter.

The necrosis brought on by infecting root systems with zoospores or inoculating decapitated plants with mycelium was diminished in the transformants, including the $K_{13}V$ mutant. The level of resistance rose when the foreign gene was in the hemizygous state, which is similar to the enhanced penetration of the T' phenotype in tobacco transformed by native Ri T-DNA and related with hemizygosity.

6.4 RESISTANCE

The levels of salicyclic acid and PR proteins remained at baseline levels in control and transgenic plants until they were tested by Phytophthora, indicating that the mechanism of improved resistance was not constitutive systematic acquired resistance (SAR) (Bi et al., 1995).

Increased secondary metabolites with antifungal properties were thought to be interfering with Phytophthora infection. Necrosis of Phytophthora would release fungal genes for elicitins like cryptogein into the soil, but they would need to be integrated into a T-DNA in order to be conveyed to plants, necessitating DNA homology between a soil bacterium like A. baylyi and Phytophthora. T-DNAs might be present in the bacterial and fungal genomes, which would provide the homologies necessary for the transfer of genes like cryptogein into A. baylyi, where Ri plasmids are maintained, since some bacteria, including Agrobacterium, can genetically transform fungi.

The elements required for the existence of a multidirectional network of genetic links between plants, fungi, and bacteria were identified, and the transfer from a fungus to a plant resulted in increased resistance to the fungal pathogen, showing it could have occurred in nature. However, since such transfers would occur on an evolutionary time scale, human intervention was required to demonstrate their feasibility in nature.

Researchers were able to identify the calystegins as glycosidase inhibitors. More knowledge had combined about the biochemicals than the enzymes that made them. Biochemicals are shared by all species, and they have developed from basic precursors to become a variety of forms and activities. Numerous dicot species preserved the changed phenotypic. The Ri T-DNA genes should logically work via conserved biochemistry.

Since polyamines are involved in plant growth (such as flowering) and are present in all living things, we can choose to investigate any relationships that might exist between changes in polyamine accumulation and the altered phenotypic(A. Kumar et al., 1997).

6.5 TOBACCO

In Versailles, researchers created tobacco plants that only expressed rolA or rolC from the Ri TL-DNA or the wild type Ri T-DNA. Additionally, they produced tobacco plants that express these genes alone when grown under the 35S-CAMV (Cauliflower Mosaic Virus) promoter. Tobacco plants exhibiting the transformed phenotype had decreased free polyamines and polyamine hydroxycinnamic acid conjugates, and that this reduction occurred in direct proportion to the severity of the phenotype: for example, there was a greater decrease in polyamine and polyamine conjugate titers in the T' phenotype than in the T phenotype. Plants with high T' phenotypes were fed polyamines, which attenuated the T phenotype. A phenotype in tobacco that resembled the transformed phenotype brought on by the Ri TL-DNA was produced by the compound DL-Difluoromethylornithine (DFMO), demonstrating that the inhibition of polyamine accumulation via the ornithine pathway is a crucial link in the chain of biochemical events that results in the transformed phenotype. rolA was the key factor in the T' phenotype.

The severe T' phenotype in 35S CAMV rolA plants prevented them from flowering. Because 35S CAMV rolA plants lacked the polyamine conjugates that accumulate in the top of the shoot before to flowering in wild type plants, this severe phenotype could not be attenuated by spraying 35S CAMV rolA plants with polyamines. Grafting P35S-rolA shoots onto wild type plants that had been driven to flower had accumulated conjugated polyamines in their branches since conjugated polyamines were unavailable for watering tests.

A wild type rootstock used as a control that had not been stimulated to flower failed to bring back flowering. In a final set of attenuation studies, some researchers found a correlation between phenotypic attenuation, a decrease in rolA transcripts, and methylation of locations in the 3' regulatory region utilizing plants producing rolA from wild type regulatory sequences. The polyamines and polyamine conjugates decreased the phenotypic changes brought on by transformation by the Ri TL-DNA and rolA alone.

6.6 T-DNA

The Ri T-DNA encodes enhanced flexibility, such as the production of roots and shoots, to be passed on through meiosis(Tamas, 1995). However, increased plasticity is not adaptive if it affects fertility. Homozygosity, segregation of truncated T-DNAs, physiological compensation (by producing more polyamines and their conjugates), and decreased transcription, for example, through methylation of the 3' regulatory region, are all methods for attenuation. Thus, regulation of the alien DNA ensures sexual transfer. The progeny from self-fertilization would be selected for through homozygosity-based attenuation, whereas the offspring from outcrossing would be counter selected. Thus, it ought to promote speciation. Since life has existed on Earth since the beginning of its history and all of it has been studied, it is clear that the living forms we are familiar with share a common ancestor. These observations support the old idea that life exists everywhere, but they do not show panspermia. Researchers examined short wavelength UV absorption in known and potential UV shields using the UV beam from the synchrotron at the University of Aarhaus, Denmark, and discovered that flavonoids showed UV absorption properties similar to those of DNA. Using mutant Arabidopsis (A. thaliana) plants deficient in sunscreens, preliminary studies at the DLR (German Aerospace Center) in Cologne, Germany, revealed that flavonoids were crucial in shielding seeds from a portion of the UV spectrum that is harmful to unprotected organisms. Arabidopsis and tobacco seeds were also found to be significantly more resilient than others. A 110 bp unprotected DNA fragment's survival was also discovered by PCR. Morning glory seeds, which have thick seed coats and are long-lived in nature, were much more resistant to UV 245 nm than Arabidopsis and tobacco seeds This finding suggests that larger seeds with more protective coats should survive much longer exposures. They examined the structure and operation of a nptII marker gene carried by the plastids in exposed tobacco seed plants because they lacked morphological mutations. In A. baylyi, we utilized quantitative PCR and marker exchange to demonstrate that DNA damage was being avoided and repaired and that other targets such as ribosomes, membranes, and proteins were more detrimental to seed.

In order to assess the DNA integrity of exposed seeds, marker exchange was used as a standard for homologous recombination in naturally transformable bacteria like A. baylyi, which is how DNA reached the biosphere. It is possible for naked or protected DNA from Earth's biosphere to be carried by updrafts into the stratosphere, return to Earth at a later time, and reenter our biosphere through homologous recombination in bacteria like A.

baylyi(Harms & Wackernagel, 2008). Fragments of the nptII gene survived unprotected exposure to full space conditions. Natural transformation of A. baylyi needs at least a few hundred base pairs of one-sided homologies between the incoming and recipient DNA in order to be effective. This limits the inward transfer of extraterrestrial DNA to sequences originating on Earth and implies that life on Earth is immune to exoDNA. On the other hand, partially degraded DNA and bacteria delivered by micrometeorites or extinct extraterrestrial species that resembled seeds may infiltrate Earth's biosphere if DNA transmission into the biosphere was not homology independent. There may be species of the mostly unstudied bacteria that are able to incorporate heterologous DNA into their genomes, for instance under stressful conditions. As an alternative, universal, homologous sequences like the Ri TL-DNA that would genetically connect dispersed biospheres could allow for exoDNA to influence DNA sequence evolution. A logical explanation for how life first appeared on Earth in the past is the transport of nucleic acids through space. The increased temperature from the Sun will, however, put a cap on our ability to live in the future. Life must be exported to exohabitats, for example, in a vector like a plant seed, in order to survive. Ironically, dispersal away from Earth has previously happened as a result of UV-resistant bacterial spores being introduced to unmanned spacecraft. As a result of our space activities, exospermia (as opposed to endospermia) is occurring, reversing the guided panspermia concept proposed to explain the beginning of life on Earth.

In the case of seeds, they might be genetically altered to accumulate more of the flavonoid sunscreens or chosen for their innate resistance(Schenke et al., 2019). They might also be covered in UV protection and packed with naturally occurring, beneficial microbes (e.g., nitrogen-fixing Rhizobium). Ploidy levels can be raised in plants to increase genetic redundancy in nuclear DNA. It may also be possible to increase the redundancy of fragile, recovery-related components including ribosomes, chloroplasts, and mitochondria. The latter two might be genetically altered using efficient DNA repair systems from bacteria, such as those found in *Deinococcus radiodurans*. Following the first in vitro culture of altered roots, various biotechnological uses were either self-evident or suggested by others. Since A. rhizogenes transformation happens in nature, it's frequently thought of as natural and doesn't entail genetic engineering. As a result, an industry that creates inventive and affordable biotechnological applications using wild type A. rhizogenes has emerged. Because it calls for more creativity, researchers feel that low budget research is frequently more innovative and adventurous than the well-funded counterpart. Additionally, it is possible in developing nations, where innovation has a promising future. A. rhizogenes is thought to create a modified, secondary root system in nature. A. rhizogenes could take hold of stubborn apple rootstocks and opines can be transferred from chimeric apple trees' genetically altered roots into their wild-type plant parts. An apple tree that was genetically altered produced more roots and shoots, as would be predicted given its greater flexibility.

Mining, manufacturing, sewage sludge, and batteries all release heavy metals. They are dispersed by animals and plants that absorb and concentrate them as well as by water, wind, and human activity. They remain in the environment after being released, where both plants and animals can easily consume them. However, heavy metals can interact with soil components and be removed from the biological system as a result. While it is relatively easy to assess the presence of heavy metals in soil, there has never been a cheap and accurate approach to determine their bioavailability, a vital factor in determining toxicity. Therefore, we set out to use altered roots as a living assay for bioavailability. The outcome was a feasibility study that demonstrated the utilization of transformed tobacco and morning glory roots in contaminated sewage sludges to measure the availability of Cd. Researchers observed that cultured roots that have been modified take up little room, are affordable to create, eliminate the complicated and changeable effects of aerial plant components, and offer simplicity and reproducibility. The goal was to collect the morning glory as it eliminated Cd from the soil while first stabilizing the location to prevent erosion. Modified morning glory roots were utilized to cultivate an arbuscular mycorrhizal (AM) fungus, an obligate rhizosphere symbiont. About 80% of the vascular plants depend on AM fungus for their growth. Invading hyphae trade energy-dense nutrients from the plant for minerals (such as phosphorus) in the root cortical cells. Hyphae build substantial soil networks, support the host's root system, and physiologically link many species of plants. Their spore production in potted plants had been attempted, but it was abandoned out of concern that it would spread infections. This issue was resolved using axenic root cultures.

6.7 TRANSGENIC RESEARCH

Both fundamental research and the development of biotechnology products depend on the effective delivery of macromolecules into plant cells and tissues. The objective of transgenic research is to transfer DNA molecules to cells that can regenerate and stabilize their integration into the genome. Numerous macromolecule delivery techniques have been researched during the past 40 years. Particle bombardment and Agrobacterium-mediated transformation are the preferred techniques for DNA delivery to create transgenic plants (Travella et al., 2005). New requirements for biomolecule delivery have been created by the quick development of genome editing technology, which has also renewed the development of new transformation methods. Many of the previously researched gene delivery methods are now being used to distribute genome editing equipment for a variety of applications. Modern DNA assembly techniques enable quick cloning of large, multigenic vectors for plant transformation in comparison to conventional methods. These techniques can combine several DNA fragments into a single vector in a single process and are frequently reliable and effective. The expression of various genes for trait stacking in the crop of interest was enabled by the varied DNA components present in these DNA constructions. Automation equipment is integrated, and high-efficiency cloning techniques are used to speed up the DNA assembly and validation process on this platform. By using this platform, bottlenecks for ordinary molecular biology are eliminated, and new opportunities for evaluating experimental ideas downstream are created.

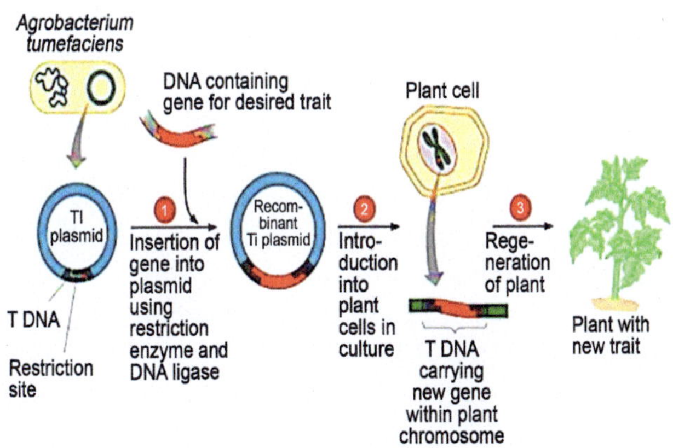

Figure 6.3. Agrobacterium-mediated transformation. (Source: https://www.researchgate.net/publication/348908271/figure/fig1/AS:985721422761989@1612025709772/Agrobacterium-mediated-plant-transformation.jpg, Creative Commons License).

It has been documented that EMT technology can successfully change both monocot and dicot plants due to its versatility and effectiveness. The modified rice and oilseed rape plants used were developed utilizing

EMT protocols that were based on common Agrobacterium-mediated transformation (AMT) (Figure 6.3) procedures to ease community users' transition to EMT(Zuniga-Soto et al., 2015). *Setaria viridis* is a developing model for C_4 species and is crucial for validating some genes for future C_4 crop transformations, including those of wheat, maize, and sugarcane. *Agrobacterium tumefaciens*-mediated stable transformation of S. viridis use reporter genes and selectable markers. Using embryogenic callus from S. viridis, routine transformation efficiency was increased to 29%. As an alternative, researchers created a floral dip transformation technique with a 0.6 percent efficiency. The suggested techniques may be helpful for genetic and genomic analyses of significant C_4 crops used for food, feed, fiber, and fuel. Rapid and transitory gene expression is the result of transient transformation or transient expression. This method is frequently employed as a first-tier screening technique for the assessment of factors that influence gene expression. Here, researchers offer a method for testing promoter elements that affect gene expression by bombarding lima bean cotyledons with particles bearing constructs encoding the green fluorescent protein (gfp) coding area. Though conceptually relatively easy, if appropriate techniques are not properly followed, this lima bean transient expression system may not operate successfully.

Despite maize transformation being a practice that has been around for more than 25 years, the technology has remained too specialized, labor-intensive, and ineffective to be of much value to academic labs. This issue is made worse by the fact that future requirements for maize genome engineering almost certainly go beyond what experts currently refer to as conventional maize transformation techniques. Scientists have created a new expression system for Bbm and Wus_2 that is an improvement over the first-generation system. This system uses a non-constitutive maize phospholipid transferase protein promoter (Pltp pro) to drive Bbm expression and a maize auxin-inducible promoter ($Axig_1$ pro) to drive WUS_2 expression. With this set of expression cassettes, zygotic immature embryos that have undergone Agrobacterium transformation quickly develop a large number of somatic embryos on their scutella. These somatic embryos undergo uniform transformation and can grow into plants without going through a callus phase. In as short as one month, transformed plants are transferred to the greenhouse; these T_0 plants are fertile and match the phenotypic of the inbred that was obtained from the seed. T_1 seeds have a uniformly wild-type inbred phenotype and germinate normally. With the help of this innovative technology, transgenic T_0 plants might possibly be produced in great quantities in B_{73}, Mo_{17}, and the recently created Fast-Flowering Mini-Maize through a quick and simple transformation process. In order to understand how agronomic features are regulated at the molecular level in rice, a model monocot cereal crop, genetic transformation is crucial. Here, we need an effective strategy for applying the Agrobacterium-mediated transformation technique to produce transgenic japonica rice seedlings. An inexperienced lab technician can simply follow the protocol because it simplifies the medium compositions and transformation steps. In the last ten years, we have changed thousands of gene constructs using this approach, and more recently, we have used the CRISPR-Cas9 system to edit hundreds of genes.

One of the most significant bioenergy crops for lignocellulose ethanol generation is switchgrass (*Panicum virgatum* L.). By adding or changing desirable genes, molecular breeding offers a potent tool to supplement traditional switchgrass breeding. Some researchers outline lowland tetraploid switchgrass cultivar Alamo transformation techniques using Agrobacterium tumefaciens. One of the most popular techniques for determining trait composition and studying gene function in wheat is biolistic transformation. It has been widely used as a basic framework to produce wheat plants with novel features and has developed into a potent tool for assisting crop development. Even though wheat is still one of the most difficult plant species to modify, this procedure provides a highly effective method for creating transgenic plants. Researchers created transgenic wheat using this methodology by co-bombarding two plasmids with the genes for the green fluorescent protein and the glufosinate herbicide selection, respectively. Both the spring and winter types of transgenic bread wheat (*Triticum aestivum*) have been produced with success. By carefully controlling the enzymatic digestion, the plant cell wall is removed without causing any harm to the cell membrane, yielding plant protoplasts. For many biochemical and physiological experiments, protoplasts serve as a real single-cell system. The utility of protoplasts is expanded from transient expression experiments to the creation of persistent transformation events by the ability to regenerate protoplasts from a variety of agriculturally significant crop species into a fruitful entire plant. The recovery of putative transgenic canola plants and key processes, such as enzymatic digestion for protoplast release, density gradient-based protoplast purification, PEG-mediated transfection, bead-type culture (sea-plaque agarose and sodium alginate), needs to be studied. This technique has been applied for routine commercial-scale production of stable transgenic canola events as well as genome editing mediated by double-stranded DNA breaks.

The efficiency of the work is considerably increased by simplifying transformation processes. Standardized Agrobacterium-mediated transformation procedure for alfalfa (Medicago sativa L., Regen SY4D) and Medicago truncatula was framed(Minz & Sharon, 2010). The medium is supplemented with cytokinins to promote shoot regeneration, and sonication is used to increase Agrobacterium infection. Alfalfa has a transformation efficiency of more than 90%, but M. truncatula, which depends on many selectable markers in the gene constructs, has a comparably lower efficiency of up to 60%.

Transgenic plants can be produced in a time frame of 4 to 8 months, with an average of 6 months. The same media types are used for both species using this unified approach, which results in considerable time and resource savings. More than 150 plant species from different plant families have transformation protocols for Agrobacterium-based transformation, biolistic bombardment, and PEG transformation of protoplasts. The leaf-disc method is the accepted procedure for Agrobacterium-mediated transformation of several Populus genotypes(Xia et al., 2023). Alternative basic transformation techniques, such as biolistic bombardment and protoplast PEG transformation, are also important. One of the most crucial elements of any plant breeding programme is the development of disease resistance. Long juvenility, a high degree of heterozygosity, polyembryony, self-incompatibility, and the abortion of reproductive organs make traditional citrus breeding methods. The application of genetic transformation, which enables the addition of a specific agronomic characteristic to a plant without otherwise affecting its phenotypic, is an intriguing alternative to the traditional breeding technique. Numerous hybrids and citrus species have been transformed by the use of *Agrobacterium tumefaciens*.

Researchers were able to transfer the Bs_2 gene into citrus using this method, and they were also able to improve transgenic Bs_2 gene-expressing lines' resistance to citrus canker. In order to minimize the disruption caused by meiosis in conventional breeding, precision breeding is a method of genetic improvement that transfers only specified features across sexually compatible species via the comparatively stable mitotic cell division pathway. The availability of the Vitis genome sequence along with highly efficient gene insertion and plant regeneration methods are factors that enable precision breeding. Leaf or floral explants are used to create embryogenic cultures. For Agrobacterium-mediated transformation, somatic embryos at the cotyledonary stage of development are utilized. Anthocyanin pigmentation and antibiotic resistance are used to select transformed embryos after co-cultivation with Agrobacterium expressing the desired genes. Afterwards, somatic embryos are borne to create modified plants that are hardened and moved into a greenhouse. PCR is used to confirm the existence of the desired genes.

The most important crop that is not a cereal is the potato, which is ranked as the fourth most important food crop in the world. Potato can be easily changed with *Agrobacterium tumefaciens*. In the last 20 years, significant advancements have been made in the tissue cultivation, transformation, and regeneration of potato. The tomato plant serves as a model plant species for numerous academic activities, including studying gene function. It is also an important food crop. In addition to the vast genetic and genomic resources for tomatoes, transformation is frequently used to facilitate these investigations. The routine transformation procedure employed in lab relies on the infection of immature cotyledon sections with *Agrobacterium tumefaciens*, and it has been used with numerous tomato genotypes. Researchers have employed CRISPR/Cas9 for genome editing, RNA interference for gene silencing, and vector systems for overexpression. Selectable marker genes that conferred resistance to kanamycin, hygromycin, and the herbicide component bialaphos were present in the vectors used to create gene constructs.

Genome engineering is a biotechnological method for carefully altering an organism's genetic code in order to change the context of an existing sequence or to develop new genetic resources, such as to produce characteristics or performance enhancements. The generation of DNA double-strand breaks (DSBs) or neighboring single-strand breaks is the primary mechanism for effective targeted genome modifications (SSBs). As a result of the fact that all living things must constantly contend with stimuli that threaten the integrity of their genetic makeup, a vast variety of DNA repair mechanisms have developed. Controlling and enhancing targeted gene changes requires a thorough understanding of the many repair processes. Nonhomologous end joining (NHEJ) and homology-dependent repair are two methods for DSB repair (HDR) (M. Liu et al., 2019). Since tiny insertions and deletions are also a common result of NHEJ-mediated repair, it can be used to specifically knock out genes, reshuffle linkage groups, or even complete chromosomes. Since a homologous template is a key need for HDR, this technique can be used to specifically integrate ectopic genes into the plant genome. Targeted DSB induction in the plant genome is made possible by the invention of various artificial site-specific nucleases. DSB repair in vivo has been studied qualitatively with regard to mechanistic variations using such synthetic nucleases, as well as statistically to identify the critical elements for NHEJ and HR, respectively. The findings of these studies support biotechnological applications for transiently altering the plant DNA repair machinery in favor of targeted genome engineering, allowing for a better understanding of genome evolution and aiding in the identification of synergistic or antagonistic genetic interactions.

CHAPTER 7

Advancements in Plant Tissue Culture

CONTENTS

7.1 INTRODUCTION

Plant tissue culture (Figure 7.1) refers to all methods for raising plant cells, tissues, or organs in a certain way in an artificial environment to provide consistent planting material in large quantities. It also discusses how to enhance the characteristics of the plant. Cuttings made from somatic cells or seeds that use zygotic cells can produce plants outside of a laboratory setting. It is significant that somatic cells have the potential to regenerate at the very least. Meristematic tissue is used in the propagation process when cuttings are used(Prammanee et al., 2011). The zygotic and somatic cells, including non-meristematic tissues, nevertheless contribute to plant propagation in in vitro settings. The main distinction between plant propagation in the field and plant propagation in a lab is in vitro conditions. Plants are multiplied in large numbers as a result of such micropropagation in a controlled setting.

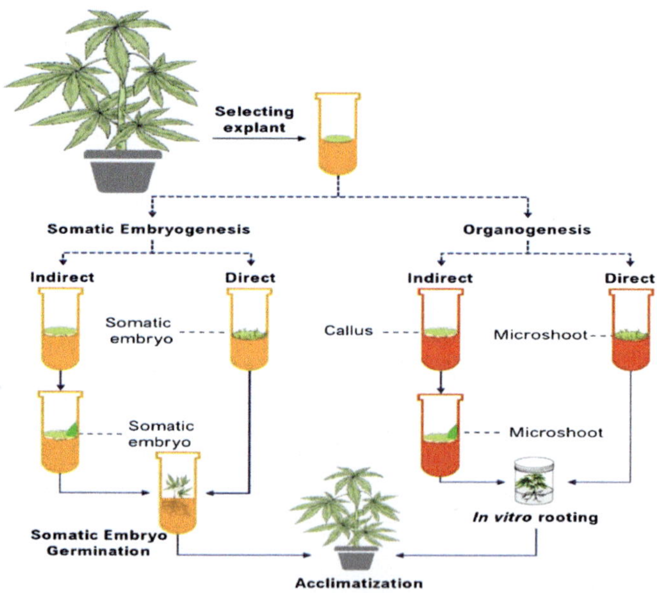

Figure 7.1. Plant tissue culture. (Source: https://www.researchgate.net/profile/Mohsen-Hesami/publication/351881072/figure/fig2/AS:1027751133851650@1622046373876/The-schematic-diagram-of-plant-tissue-culture-procedures_W640.jpg, Creative Commons License).

7.2 AGRICULTURE

In fact, half of Brazil's sugarcane (Figure 7.2) is used to produce ethanol, a sustainable green fuel, giving sugarcane a strategic place in global agriculture and status as the main crop for renewable energy.

Figure 7.2. sugarcane. (Source: https://www.researchgate.net/publication/337180222/figure/fig1/AS:840043203682304@1577293315050/An-energy-cane-RB-hybrid-and-a-sugarcane-variety-SP791011-at-six-months-after-planting_W640.jpg, Creative Commons License).

As the most crucial basic input and the only one capable of improving crop productivity by 15% to 20%, sustained production and supply of high-quality seed is crucial for the stability of agricultural production. The constant supply of sufficient amounts of high-quality seed cane, which must be genetically pure, free of diseases and pests, and without nutritional abnormalities, is the primary determinant of the maintained high output of sugar per unit area. Since the produced plants are devoid of viruses and retain the original vigour of the newly bred variety, this can only be accomplished via tissue culture procedures. As the creation of many plantlets from the explant of a single plant/plants of a single genotype using in vitro culture of the explant, micropropagation is a potent tool. The sugarcane micropropagation technology is well known and has demonstrated a thousand-fold rate of multiplication. At the moment, it is one of the fastest seed production techniques accessible, offering a speedy method of producing viable seed of novel types. The technology is not only profitable but also economically viable.

7.3 TISSUE CULTURE

One of the most well-known approaches is plant tissue culture, which has been used since 1902 to create disease-free plants with the appropriate quality at a higher output rate over a shorter period of time, regardless of the season(Twaij et al., 2020). Each plant has its own unique form and dietary needs. Additionally, different tissues in different parts of the plant have different nutritional needs. The selection of the appropriate media and its composition are therefore essential for the best possible growth of the target plant. The development of appropriate plant tissue culture techniques depends critically on the formulation of the nutritional medium for the specific tissue and the individual plant. Macroelements, microelements, nitrogen sources (including amino acids), carbon sources, organic supplements (primarily vitamins and other supplements), growth regulators like auxins, cytokinins, gibberellins, abscisic acid, ethylene, and solidifying agents, are the main components of the plant tissue culture media. A specific concentration of the fundamental components must be incorporated. For organogenesis, plant growth regulators are added, and the balance between them relies on the type of tissue culture technique used(Gaspar et al., 1996). Activated charcoal, antioxidants, ethylene inhibitors, complex organic additives, and antibiotics/antifungal agents are some of the additional supplements that are included into the media to accomplish certain effects. These supplements also reduce plant loss from contamination. Plant growth regulators (PGRs) are substances that plants employ to support their development, differentiation, and growth. Plant tissue culture has gained popularity as an efficient method of in vitro plant propagation since it was first used in the early 20th century. Tissue culture is employed in studies on plant developmental processes, from micropropagation to the creation of transgenic plants, among many other applications, despite all the challenges. Synthetic PGRs are being introduced to the basket along with natural PGRs to enable the in vitro growth of different species. In addition to commonly utilized auxins and cytokinins, newly defined PGRs, such as salicylic acid, jasmonic acid, and brassinosteroids, are presenting fresh opportunities for plant regeneration, including suspension cultures.

7.4 STRATEGIES

Improved strategies for growing recalcitrant species will be made possible by novel techniques for characterising endogenous PGR levels with recently discovered PGRs(Roberts & Hooley, 1988).

Figure 7.3. Gottlieb Haberlandt. (Source: https://upload.wikimedia.org/wikipedia/commons/thumb/9/9f/Gottlieb_Haberlandt.jpg/800px-Gottlieb_Haberlandt.jpg, Creative Commons License).

Gottlieb Haberlandt (Figure 8.3) initially used the term totipotency in 1902, and now, everyone working in the field of plant tissue culture is well familiar with it. The technology of plant tissue culture is currently widely employed for micropropagation, conservation, and transformation. Plant cells can differentiate, dedifferentiate, and redifferentiate numerous times, demonstrating their plasticity. The plant's reproductive and somatic cells both exhibit this developmental adaptability. Two major mechanisms of propagation/regeneration exist, namely somatic embryogenesis and organogenesis in plant tissue culture, which take advantage of the totipotency and plasticity properties of the plant cell. A single somatic cell that eventually regenerates into an entire plant is used in the process of somatic embryogenesis, which results in the formation of an embryo. Somatic embryos have a bipolar structure with both apical and basal meristems that can develop into a shoot or a root, respectively.

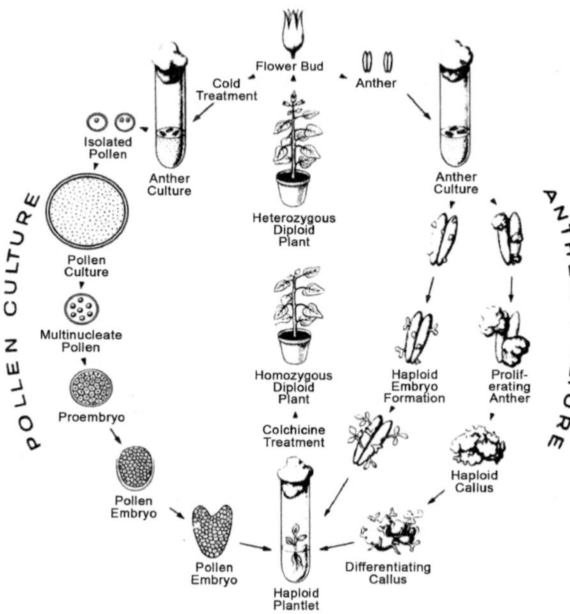

Figure 7.4. Haploid plants. (Source: https://upload.wikimedia.org/wikipedia/commons/thumb/9/9f/Gottlieb_Haberlandt.jpg/800px-Gottlieb_Haberlandt.jpg, Creative Commons License).

Organogenesis is the process of creating organs from explant material, including the shoot, root, flower, and tuber. Despite the repeated reports of these basic plant regeneration pathways, a thorough understanding of their subtypes, influencing variables, benefits, limits, and applications in crop improvement is still lacking.

The genetic transformation technique has advanced significantly over the past ten years and is now commonly utilized for enhancing the quality and productivity of superior plants as well as for scientific research into the function of a certain gene in plants. It is strongly advised to use tissue culture methods for genetic modification. In plant tissue culture, it's crucial to consider the ideal nutrient media and in vitro circumstances to support plants' healthy growth and development following genetic modification. Plant cells that have received foreign genes directly or indirectly through the use of Agrobacterium-mediated plant transformation will express the genes and create transgenic plants with the desired enhanced characteristics. Many essential genes are widely introduced into plants for a variety of functions, including micropropagation (orchids, rice, eucalyptus), enhancing pest and pathogen resistance (citrus, wheat, tomato, rice), boosting the production of nutraceuticals and pharmaceuticals (tomato), and speeding up flowering (rice, orchids). In the context of contemporary biotechnology, genetic transformation is an excellent tool because these features are crucial to the agriculture industry.

7.5 ANDROGENESIS

Androgenesis is the process of producing haploid cells from anther or pollen cultures (Figure 8.4). The sporophytes with the highest number of gametic chromosomes are known as haploid plants. Haploids are now a crucial resource for mutant breeding and the screening of desirable features. In vitro anther and isolated microspore culture are the most efficient and popular haploid production techniques. By inducing or allowing spontaneous chromosomal duplication of the haploids during a different culture, doubled haploids (DHs) can be produced (Germanà, 2011).

The use of haploid technology and the creation of homozygous plants has had a huge impact on agricultural systems. Haploid production is a very effective method. By enabling the development of fully homozygous plants in a single step, haploid and double haploid plants shorten the six to seven generations needed for conventional breeding techniques. A number of agronomically important plants now have breeding programmes that include the production of haploids by anther culture.

However, so far, 170 or so species have been examined for the formation of haploids using another culture. Therefore, it is necessary to make use of this approach in order to develop more economically significant haploid plants. The leaf blades of many species and cultivars of valuable fruit, ornamental, and essential oil plants have been studied histologically. On modified MS culture medium, plants are cultivated for varying lengths of time at ideal temperatures, light levels, and photoperiods. The structural characteristics of the epidermis, mesophyll, vascular, and mechanical components are determined by their cross-sections. The morphology of trichomes, including tissue thickness, stomata organization type, and estimation of their size and number are also important. The palisade ratio is estimated for plantlets that have distinct mesophyll.

For the maintenance of plant biodiversity, domesticated crops and wild species must be managed and protected. An alternate and beneficial option for propagating and conserving a variety of plant species, particularly vegetatively propagated plants, plants with resistant seeds, and plant species at risk of extinction, is provided by plant tissue culture-based in vitro culture techniques. The technique of plant tissue culture enables the quick, medium-term, or long-term propagation of such plant species with high rates of multiplication. The slow growth culture technique is used to preserve plant germplasm for short- to medium-term periods under growth-restrictive circumstances. By lengthening subculture intervals, germplasm storage can be extended from a few months to 1-2 years. The techniques used to save plant germplasm for slow growth include incubating cultures at low temperatures, using limited growth media by reducing the amount of nutrients in the medium, or using growth inhibitors or osmoticum to restrict culture growth.

The best and most practical method for the long-term preservation of plant genetic resources is cryopreservation. The applicability of cryopreservation for the long-term preservation of plant genetic resources has risen due to technological advancement, particularly recently created droplet-vitrification and cryo-plate technologies. Instead of having a number of restrictions, a large variety of plant species are now regularly preserved using both slow growth storage and cryopreservation techniques. Human health depends on medicinal plants. Since ancient times, they have been widely utilized in conventional medical systems all throughout the world. They are employed in drug development projects and are a valuable source of natural ingredients. People take these plants out of the wild and use them for commercial purposes. As a result, several important therapeutic plants have become extinct. Somatic embryogenesis is the process by which we create embryos that become full-grown plants(Pachota & Orłowska, 2022). Depending on the optimization of the media, growth regulators, and other parameters, these somatic embryos can be produced either directly on the explants or through callus. These somatic embryos are useful for genetic modification, secondary metabolite generation, micropropagation, and conservation. Various medicinal plants have been reported to undergo somatic embryogenesis.

The enhancement of crops has always been a top priority in order to feed the world's expanding population. The basis of all breeding strategies is the flowering features of plants since they cause genetic rearrangement. Flowers can produce seeds after being pollinated naturally or artificially, which enables conventional breeding techniques to alter their genetic makeup. Any crop eventually reaches a point where its genetic potential is depleted and it can no longer be enhanced. Breeders are seeking for wild relatives of the crop in order to find candidate genes that have desired features. Genetic incompatibility and environmental variables can still cause an embryo to abort even after successful fertilization. At the molecular level, postfertilization transcription factors inhibit embryo growth and obstruct the generation of viable seeds. Technically known as embryo rescue, interspecific hybridized embryos are cut out in vitro and cultured on media to save them from being aborted. One of the earliest in vitro plant breeding techniques is ER. The genotype of the crop, the culture medium, plant growth regulators, the age of the fruit, and the later developmental stages of the saved embryos are only a few of the variables that affect how well the ER approach works. The torpedo stage embryo has the highest success rate when compared to other embryonic stages. This method can be used to address a number of causes of interspecific hybridization failure, including premature germination, malnutrition, cytological alterations during embryogenesis, endosperm number, and polar nuclei activation. ER is useful for studying seed physiology, viability, and dormancy as well as ploidy and different chromosomal elimination studies. As a result, this technique has proven to be particularly successful for the generation of interspecific hybrids in a variety of agricultural, ornamental, and vegetable crops. The ER approach makes producers, plant breeders, and the seed business more able to ensure the safety of the world's food supply.

7.6 REGENERATION

A potent technique for both plant production and trait improvement are plant tissue culture. Regeneration, whether through meristematic tissue or differentiated tissues, is the first step necessary for subsequent physiological processes in both circumstances. To achieve the required organogenesis, a variety of plant growth regulators, medium, physical variables, plant age, explant type, and season are handled. Despite these alterations, several plant species that were very resistant for tissue culture purposes failed to start the organogenesis process. Loss of totipotency, loss of morphogenetic competence, and problems germinating are all examples of in vitro plant recalcitrance(Benson, 2000). Regenerations are greatly influenced by stress-induced free radical production and the condition of endogenous plant hormone concentrations. Crop improvement can be achieved by genetic transformation, which involves modifying the plant's cellular DNA using recombinant DNA techniques. Plant cellular genome can be genetically altered using a variety of biological, physical, and chemical techniques. The transformation process can also be aided by techniques for in vitro plant tissue culture or a few in vivo techniques. The genotype specificities, treatment circumstances, Agrobacterium strains used, immature and juvenile explant sources, explant physiology, etc. all affect how well the genetic transformation goes. Being a natural genetic engineer, the soil-borne gram-negative bacterium Agrobacterium spp. has been used to genetically modify numerous plant species. While chemical approaches manipulate transgenesis by using chemical mediators, physical procedures use particle bombardment/biolistic method, sonication, microinjection, macroinjection, electroporation, nanoparticles, and silicon carbide fiber-mediated transformation. In addition to the use of synthetic promoters, transcriptional activators and repressors, and tools for the assembly, synthesis, and transformation of large DNA inserts to achieve high frequency and precise genetic manipulation, the success in the generation of transgenic plants paved the way for advancements such as multigene transfer, site-specific integration, and regulated gene expression. Genome editing technology has also used genetic transformation techniques, and crops that have had their genomes altered with great success. Modern technology should encourage the development of genetically enhanced plants with superior economic and commercial features for use in industry and agriculture.

Because they are sessile and lack a well-developed immune system, plants are constantly under biotic stress from their surroundings, including viruses, bacteria, nematodes, fungi, and insects. It's possible that the bacteria, fungus, nematodes, and insects are restricted to a particular time frame for the plants. In addition, once viruses have integrated into a plant's genome, they become a permanent component of the plant and continue to harm it indefinitely. Because there are no effective treatments for viruses, they pose one of the greatest threats to plants and may result in quality and production loss. Plants that reproduce asexually, such as ornamentals, ornamental bananas, potatoes and sugar cane, have more severe losses. The fastest-growing tissue of a plant is grown in a process called meristem culture, which is highly useful for creating plants devoid of viruses since meristem development always outpaces virus cell-to-cell transmission. Since its invention in 1952, the method of employing meristems to produce virus-free plants has become widely employed to get rid of plant viruses from several economically significant crops. It has been demonstrated that many plants respond better to the meristem culture when combined with heat, chemicals, and radiation to eradicate viruses. The genetic stability and development capacity of meristem tissue are higher. Plant tissue culture is the aseptic and controlled growth of cells, tissues, and organs on nutritional media. It can be used for a variety of things, including plant mass multiplication, germplasm preservation, the creation of pathogen-free plants, the manufacture of secondary metabolites, and the growth of transgenic plants. Understanding fundamental and practical areas of plant science, such as embryology, morphogenesis, cytology, pathology, physiology, and biochemistry, is made much easier by this. An effective plant regeneration technique is a crucial necessity for the genetic modification of plants. Screening and choosing superior somaclonal and gametoclonal variations have made it possible to improve a number of crops. Commercial cultivation of such plants offers a number of advantages over traditional breeding procedures since it requires less time and work.

Tissue culture is employed with medicinal plants to effectively produce active chemicals for the pharmaceutical industries(Yue et al., 2016). Due to its therapeutic capabilities against microbial infection, inflammation, and cancer, medicinal plants have found various uses in the pharmaceutical industry. Several studies have been done on them for their preservation and effective generation of secondary metabolites. The primary disadvantage of this method is that all economically significant plants cannot survive in an environment other than their natural habitat. In addition, the plants are also exposed to pathogen attack and other stress factors that occur in the environment. Traditionally, the production of secondary metabolites has been done by the massive cultivation of plants, which requires more land. So, the problems with the generation of secondary metabolites have been solved through tissue culture techniques. The use of natural remedies is widespread and keeps growing. Because they are taken from wild environments, genetic diversity is lost. Domestic cultivation is a viable alternative that offers a chance to get over some of the issues that come with herbal extracts naturally, such as the enhancement of genetic and phenotypic variability, misidentification, extract variability and instability, poisonous components, and contaminants. Plant tissue cultures were carried out for the efficient manufacture of pharmaceutically significant chemicals as well as for the preservation of medicinal natural resources. Utilizing various growth media and cryopreserving the sterile in vitro plants, in vitro organ culture can be employed as an alternative technique for the conservation of endangered genotypes. Tissue culture aids in maintaining the genetic makeup of plants and helps prevent the loss of the conserved heritage owing to biotic or abiotic environmental changes and pressures. Many plants with dormant seeds or plants without seeds cannot be kept for an extended period of time. In vitro methods can be used to conserve such species in order to maintain their gene banks. Thus, tissue culture is essential for the preservation of medicinal plants that are in danger of extinction as well as serving as a reliable supply of material for biotechnological research projects.

7.7 APPLICATIONS

Since ancient times, pharmaceuticals/drugs made from plants and herbal remedies have been used to cure human and animal diseases, and many countries still rely heavily on plants and herbs as their primary drug supplies today. For the synthesis of phytochemicals, plants are regarded as biofactories. A more affordable, environmentally friendly, and superior alternative to manufactured medications is phytomedicine. Many plant scientists are interested in studying these phytochemicals because of their little or nonexistent adverse effects. Tissue culture technique is now regarded as a highly effective approach for the regulated generation of such phytochemicals. Many secondary metabolites, including polyphenols, alkaloids, terpenoids, flavonoids, and saponins, are abundant in plants and are widely used to prevent human diseases. The oleaginous yeast *Yarrowia lipolytica* has emerged as an industrially relevant framework to produce various valuable chemicals(Poorinmohammad & Kerkhoven, 2021). Metabolic engineering of Y. lipolytica relies on the availability of genetic engineering tools. Existing engineering strategies for this yeast include homologous recombination, random integration, and episomal plasmid-based gene expression. CRISPR-Cas9 based genome-editing toolbox has also been developed to facilitate multiplexed gene disruption and regulation. Alternative to Cas9, the CRISPR effector Cas12a has also been adopted to perform genome engineering in multiple species. Due to its distinctive features such as short and simple crRNA structure, the ability to process its own crRNA and T-rich PAM sequence (TTTN), Cas12a holds promising potential to be developed as an efficient genome-editing tool. The delivery of AsCas12a and crRNA expression via a single plasmid is important. CRISPR-Cas12a-based genome editing could expand the genetic toolbox of Y. lipolytica, which is complementary to the classical Cas9-based tools.

Plants preserve the physical and biochemical traits that are intrinsic to a particular genotype under precise culture and preservation conditions, but they also exhibit some novel traits. Researchers found that changes in a cultivated plant's ecological and anatomical variability are primarily quantitative in nature. The range of changes at the anatomical and submicroscopic level of plant organization in vitro can be determined using the amplitude and pace of expression of the emerging structural features, revealing the dependence of the growth processes delay on the structure of regenerants. In order for in vitro generated microshoots to maintain their vitality and adapt to the post septic culture conditions, research sought to uncover various histological characteristics in their leaves. After growing for 2–14 months in photo-capsules in vitro, samples of the researched species and cultivars of several ornamental, aromatic, and fruit crops were taken. *Rosa Damascena Mill., Rosaceae Juss.* is a hybrid of Rosa damascene, Rosa gallica, and Rosa alba, as are its cultivars. They are raised in order to make rose essential oil. This is frequently utilized in the food and pharmaceutical industries as well as for the manufacture of high-end perfumes and cosmetics. The leaf plate's edge is ragged, and the leaves were trifoliate with pinnate venation. Every micro-shoot had six or seven leaves on average. Therefore, the Lany cultivar was found to have the biggest leaves and the greatest leaf shape index. The amount of light energy absorbed is characterized by this leaf shape index. It is understood that the genotype with a lower index will utilize the necessary illumination for photosynthetic processes more successfully.

The aseptic cultured plants under study also had different leaf colors; the Raduga cultivar had dark green leaves free of chlorosis and necrotic alterations. On the other hand, cultivars Festivalnaya and Lany have dark red pigmentation along the lateral edges of the leaf blades (Mitrofanova et al., 2022). When the anatomical characteristics of leaves in in vitro regenerants were examined, the following results were found: thin, hypostomatic, single-layer epidermis-covered, bifacial type leaf blades, and anomocytic type stomatal apparatus. Additionally, the adaxial epidermis cells had a sinuous form and were more prominent than the abaxial one. A single cell layer comprised the palisade mesophyll. Its thickness varied by genotype. Chloroplasts were found diffusely or close to walls, and they had an oval form. It was computed how many chloroplasts there were on average in each palisade cell. The cultivars Raduga and Festivalnaya and the cultivar "Lany" (showed notable variations based on this indicator. A larger content of plastids can be viewed as a sign of a powerful photosynthetic system that delivers significant productivity in plants of these cultivars as the quantity of chloroplasts in a cell determines the ability for photosynthesis. Numerous physiologically active compounds utilized in medicine, fragrance, and the food sector are derived from lavender (*Lavandula sp., Lamiaceae*).

Widely employed in decorative gardening are lavender species with great commercial values, such as narrow-leaved lavender (*Lavandula Angustifolia Mill.*) and lavandin (*Lavandula intermedia Emeric ex Loiseleur*). Large-scale propagation and conservation of old valuable and new cultivars and breeding lines are all done using biotechnological technologies.

The initiation and multiplication stages of in vitro culture revealed morphological variations between cultivars in narrow-leaved lavender and lavender regenerants. The lavender cultivar Belyanka produced more microshoots (up to five), which had the tiniest leaves. One or two microshoots were produced in lavender cultivars, with Rabat having a denser leaf arrangement and having lengthened internodes. The leaves of lavender cultivars have a broader size and a less pointed apex than those of narrow-leaved lavender. In addition, its leaf blades are four to five times longer than they are wide. These cultivars grown in open fields typically have leaves with similar morphological characteristics. All cultivars developed two to four microshoots per explant under the more prolonged cultivation. At the second and third pair of leaves, they split. The lavender cultivar Rabat had the biggest leaves at all stages of development. Leaf blades had a distinct mesophyll and were bifurcated. Long-term cultivation improved the leaf thickness. In the cultivars under study, the epidermis on the abaxial leaf side was composed of smaller cells. Collateral vascular bundles were found in the middle layers of the mesophyll of leaves. The lateral vein bundles contained phloem and three to four xylem vessels. The leaves were amphistomatic, and after extensive culture, it was discovered that there were significantly more stomata on the abaxial side. The stomata were raised above the bases of the epidermal cells, varied in size, and anomocytic. It should be mentioned that the cultivars under study showed significant variation in stomata distribution density. Thus, cultivars of narrow-leaved lavender had 102-180 stomata/mm^2 that were distributed nearly equally.

7.8 CULTIVARS

The 'Record' cultivar was distinguished by the smallest, most numerous stomata on the adaxial side and the largest, fewest in number stomata on the abaxial side(Chatterjee et al., 2020). As a result, this cultivar has the greatest quantity of big stomata. Oval guard cells surrounded stomata. The leaves were pubescent with two- to four-celled capitate glandular hairs and one- or two-celled simple, weakly branching trichomes. On the abaxial side, they were more numerous as well. Researchers discovered that the proportion of glandular hairs in leaf cross-sections ranged from 23 to 38 percent of the total number of trichomes. *Lavandula vera L.* cultivars have fewer than 30% of their cells with secretory activity. Researchers demonstrated that the capitate glandular hairs were where the scattered essential oil accumulated the most. The vacuoles of the mesophyll cells also produced some essential oil droplets. They can be seen in the palisade parenchyma cells. Compared to the other cultivars under study, essential oil droplets were more abundant and substantial in the lavender cultivar Rabat. Their development was observed in these cells of the bundle sheath cultivar as well. The multilayered mesophyll was made up of two or three rows of spongy tissue with a loose assembly of cells that were in contact with one another via protrusions, and one or two rows of palisade chlorenchyma with elongated, closely spaced cells. After the first passage, the palisade tissue in regenerants' leaves was a single layer that ranged in thickness from 45 to 58 m in all the cultivars that were examined. The cells were ovoid in shape. Palisade cells were seen on the leaf sections in the fourth passage, typically in two layers. The "Rabat" cultivar has the most developed palisade chlorenchyma. Palisade index rose to between 0.56 and 0.58. Interspecific hybrids have an intermediate kind of inheritance of the original species anatomical structure traits. This behavior can be explained by an increase in homeostasis, which enables them to more effectively adjust to both low and high air and substrate humidity levels. Research revealed the described morphological characteristics of vegetative organs, variations in leaf blade thickness, a number of stomata on the abaxial side, and essential oil location.

In the cultivars "Rabat" and "Record," the formation of palisade leaf tissue, cutinized integumentary tissues, and profuse pubescence are xeromorphic traits that suggest the ability to respond to water stress under ex vitro adaptation. At the same time, the big cells and loose mesophyll in the cultivars "Belyanka" and "Snezhniy bars" show the capacity to develop normally in both humid and xeromorphic environments. Additionally, the essential oil is disseminated and deposited in glandular hairs as well as in the vacuoles of the mechanical tissues and leaf mesophyll. A perennial ornamental herb in the family Cannaceae Juss and order Zingiberales Nakai is the canna lily (*Canna hybrida Hort.*). Garden canna plants were grown in vitro and took up the entire volume of the culture jars. They varied in height from 8.6 to 16.0 cm, with roots that ranged from 1.2 to 5.4 cm, and had a lot of straight or linear-lanceolate leaves. The "Suevia" canna cultivar plants possessed the biggest leaf blades. Low leaf shape indices (0.14 to 0.16) were present in all of the cultivars under study, indicating good light utilization.

The hydromorphic structure of garden canna leaves was discovered to consist of narrow mesophyll (86–111 m), developed aerenchyma and airways, and thin integumentary tissues (10–17 m). It has been observed that a single meristemoid can give rise to one to five microshoots. The lower leaves, primarily the lowest pair, were totally or partially necrotic while the upper leaves were a usual shade of green. The stomatal system was of the paracytic variety.

The mesophyll was not divided into palisade and spongy tissue; the leaf blades were bifacial and amphistomatic. Isodiametric parenchymal chlorophyll cells arranged in three or four layers made up homogeneous mesophyll. Most of the stomata were open. The "Suevia" cultivar possessed the smallest stomatal organs. The investigation found considerable changes in how they were arranged, with more stomatal apparatuses on the abaxial side in the "Suevia" cultivar.

The opposite was true for the "Livadia" cultivar, which had more stomata on the adaxial side of the leaf blade. The cells of the epidermal tissues were also homogeneous, the cells of the "Suevia" cultivar were elongated-oblong, and the cells of the adaxial epidermis were rounded-oblong. The chlorenchyma differentiation into palisade and spongy tissue took place in the leaf blades of the garden canna regenerants during the stage of adaption ex vitro. The leaves had an advanced aerenchyma in vivo and were amphistomatic. The rooted plants of the investigated garden canna cultivars had developed mechanical tissues, thickened epidermal tissues, and an increase in the quantity and arrangement of palisade mesophyll cells. Stomata on the adaxial side of the epidermis increased in number and shrank as epidermal cells grew longer. It is feasible to choose a stable and well-adapting ex vitro garden canna cultivar called "Livadia" from among the examined cultivars by taking into account the structural changes in the tissues of regenerants under aseptic and regulated circumstances of adaptation in vivo. A crucial component of commercial floriculture in Western Europe, Japan, Russia, the US, and other nations is the garden chrysanthemum (*Chrysanthemum morifolium Ramat.*). Several chrysanthemum species are extensively employed in agriculture, the pharmaceutical industry, and cooking. After 5 weeks of in vitro culture, a structural and functional examination of garden chrysanthemum microshoots was conducted. In the cultivars "Antonov" and "Krasky Oseni," the breeding form "Tango," microshoot length was 6.8 cm, and each microshoot produced 7–15 leaves. The breeding form "Tango" and cultivar "Egyptyanka," which, like cultivar "Krasky Oseni," were known for having small leaves, had the most leaves per microshoot.

The leaf blades resembled that ex-situ in form, being rounded or obovate. Chlorotic or necrotic leaves were few; they made up no more than 15% of the microshoots under study. With the exception of the "Egyptyanka" cultivar, active rhizogenesis was seen in all cultivars. "Kvartet" cultivar was determined to have the lengthiest roots. In vitro garden chrysanthemum leaf blades were amphistomatic, bifacial, and thin; the "Kvartet" cultivar's minimal leaf thickness was established. There was only one cell layer in the epidermis. Adaxial epidermis cells exhibited thicker cell walls. The cells have an amoeboid morphology. The stomatal system was of the anomocytic variety. It was discovered that between 83 percent and 98 percent of the stomata were open using the epidermal tissues that were created right after the opening of the culture arteries. The abaxial epidermis' cells were shaped like amoebae. With the exception of the "Krasky Oseni" cultivar, they were larger than the adaxial side cells in all cultivars. On the abaxial side of the leaf blades, there were more stomata. The "Kvartet" and "Tango" cultivars' plants were distinguished by their equally spaced stomata on both leaf surfaces. On the abaxial side of the leaf blades, there were numerous four- and seven-celled multicellular trichomes. The two upper cells of the T-shaped hairs were pointed radially. The hairs typically overlap as a result of this morphology. These characteristics can be thought of as the structures that limit excessive water evaporation, enabling regenerants to adapt to post aseptic circumstances more effectively.

The intercellular gaps were extensive and the leaf mesophyll had two or three, seldom four, layers(Williamson et al., 1995). The cultivars "Antonov," "Krasky Oseni," and "Egyptyanka" showed differentiation into palisade and spongy chlorenchyma. Palisade tissue had one layer of cells, whereas spongy tissue had two or three. The regenerants of these cultivars have a 0.38–0.42 palisade index. In vitro findings describe the anatomy of the cultivars under study, are consistent with in vivo findings made about the chrysanthemum cultivar "Lillian Hoek". The cultivars "Antonov" and "Krasky Oseni" are the most promising for in vitro propagation and subsequent adaptation ex vitro and in vivo when taking into account indicators like the active frequency of rhizogenesis, significant leaf thickness, structural characteristics of chlorenchyma, stomata, and trichomes arrangement. The species and cultivars of Clematis L. are used extensively in ornamental gardening and have numerous beneficial characteristics. Creating and maintaining in vitro collections of clematis cultivars as well as greatly speeding up the large-scale generation of pathogen-free plants are both made possible by the application of biotechnological methods in plant propagation. Numerous general and cultivar-specific characteristics for clematis plants in vitro were revealed through comparative examination of the morphological and histological traits of the examined clematis (Clematis L.) cultivars. The regenerants' leaf blades are still shaped like their original mother plants' cultivars. Assimilation tissues in the cultivars "Nikitsky Rosovyi" and "Alpinist" grown under aseptic circumstances had the least damage from necrosis. However, there were examples of undifferentiated chlorenchyma in the cultivars "Madame Julia Correvon" and "Nikitsky Rosovyi." The majority of the investigated leaf blades were characterized by a bifacial hypostomatic structure. The cultivars "Alpinist" and "Madame Julia Correvon" had the most variable leaf thickness, which ranged from 83 to 182 m(Mitrofanova et al., 2022).

All of the cultivars contained single-layered palisade tissue, while the leaf mesophyll had three to four layers. In this tissue as well as the spongy chlorenchyma, the cultivars "Madame Julia Correvon" and "Crystal Fountain" had the largest intercellular gaps. The highest in the cultivars "Nikitsky Rosovyi" and "Alpinist" had a palisade index of 0.29 to 0.37. The kind of vascular bundles was closed collateral. The varieties' circulatory systems developed similarly, with few differences. The epidermis was made up of a single cell layer that ranged in thickness from 11 to 21 m. The cuticle was absent. One, two, and five-celled simple trichomes were discovered during the examination (on the adaxial and abaxial sides). The trichomes ranged in length from 84 to 143 m. There were 2–16 trichomes per 1 mm^2 of the surface. The stomatal apparatuses were of the anomocytic type and either slightly elevated above or at the same level as the epidermal cells. Each cultivar's stomatal hole size ranged from 27 to 51 m, with V = 26 to 59 percent; the "Alpinist" cultivar's stomatal pore size being the smallest. Stomata arrangement had unique characteristics that were discovered; the cultivars "Alpinist" and "Madame Julia Correvon" had the most of them. The number of open stomata exceeded 60%. The tiniest epidermal cells were found in the cultivars "Alpinist" and "Nikitsky Rosovyi," which had irregular amoeboid form and were sinuous. ISE is mostly induced in grains, legumes, vegetables, ornamental, and medicinal plants using the synthetic auxin, auxin 2,4-D. In plant tissue, 2,4-D at high concentrations produces embryogenic callogenesis. The synthetic auxin dicamba (3,6-dichloro-o-anisic acid) is also used to trigger embryogenic callusing. When combined with cytokinin, auxins such as picloram, IAA, and NAA have been shown to be powerful enough to initiate somatic embryo induction in a variety of plant tissues.

Cell division, the generation of somatic embryogenesis (when combined with auxin), germination, and the development of shoots are all stimulated by cytokinin. BA (6-benzyladenine) and metatopolin are known to promote somatic embryo germination and shoot development in Carica papaya L. For the induction of somatic embryos in the Corydalis yanhusuo species, BAP and NAA work well together. Plant development processes include fruit maturation, the creation of hairy roots, and leaf abscission all involve ethylene. Somatic embryo induction and the formation of embryogenic calluses are influenced by ethylene generation during SE. *Leucojum aestivum L.* mature calli were observed to synthesize and release the endogenous ethylene hormone. An ethylene precursor is ACC (1-aminocyclopropane-1-carboxylic acid). Comparatively to the control, adding ACC (10 M) to the medium boosted ethylene production and decreased callus growth, but adding ACC (10 M) to the multiplication media promoted the development of globular-shaped somatic embryos to torpedo. Ethylene is inhibited by potassium permanganate and silver nitrate. The callus growth was accelerated by the inhibitor addition to the callus multiplication medium. For the induction, maturation, and germination of somatic embryos, equilibrium between endogenous and exogenous plant hormone levels (auxin, cytokinin, abscisic acid [ABA], and gibberellic acid) is required.

In tissue culture procedures, less differentiated or immature plant tissues are preferred. Given that the tissue already has preembryogenic determined cells, often known as embryogenic competent cells (PEDCs). During in vitro plant regeneration by SE, Murashige and Skoog (MS) Medium, Woody Plant Medium (WPM), Gamborg (B5) Medium, and Schenck and Hildebrand (SH) Medium and their modified media are utilized(Faisal et al., 2012). The frequency of somatic embryos and in vitro regeneration were affected by the medium's pH. The primary carbon source for SE is sucrose. Additionally, there are reports of the utilization of sorbitol, maltose, and galactose as efficient carbon sources in SE. SE is induced by a high sucrose content. In media formulations, nitrogen sources include potassium nitrate (KNO_3) and ammonium nitrate (NH_4NO_3). It was found that Triticum aestivum's DSE was improved by ammonium nitrate. Different nitrogen sources have been discovered to have an impact on SE in wheat. The SEs are supported by media that have additives such casein hydrolysate and AC added. The formation of effective somatic embryos is influenced by the amino acids proline, alanine, and glutamine. Callus production is inhibited by silver nitrate, which also encourages effective somatic embryo development. Activated charcoal (AC) is a porous carbon-based substance with no flavor that is organized in a quasi-graphitic fashion. It has a distinctive adsorption quality. It absorbs harmful metabolites that are released in an in vitro culture medium, such as 5-hydroxymethyl-furfural. During sterilization, the breakdown of sucrose occurs at 121°C, producing the metabolite 5-hydroxymethyl-furfural. This boosts somatic embryo induction, somatic embryo development, and somatic embryo germination and rooting frequency.

Date palm somatic embryo development is increased by around two times when AC is supplemented alone, but nearly three times when AC and sucrose are combined. *Cinnamomum camphora L.* produces more somatic embryos per immature zygote in the presence of 1 g L 1 AC (16.29) compared to the control (12.96). AC increases the incidence of somatic embryo germination and rooting. Even though the explants and media utilized are the same, different genotypes of plant species respond differently. Under in vitro circumstances, the morphogenesis of plants is also influenced by the frequency, wavelength, and duration of illumination exposure. Different plant tissues exhibit noticeably better embryogenic callus development in the dark.

References

1. Abdulbaki, A., Alsamadany, H., Alzahrani, Y., & Olayinka, B. (2022). Rubisco and abiotic stresses in plants: Current assessment. Turkish Journal of Botany, 46(6), 541–552. https://doi.org/10.55730/1300-008X.2730
2. Abebaw, Y. M., Tobiaw, D. C., Abate, B. A., Eshete, B. K., Seymour, S. K., & Tesfaye, K. (2021). Plant Tissue Culture Research and Development in Ethiopia: A Case Study on Current Status, Opportunities, and Challenges. Advances in Agriculture, 2021, e9979549. https://doi.org/10.1155/2021/9979549
3. Adams, S. R., Cockshull, K. E., & Cave, C. R. J. (2001). Effect of Temperature on the Growth and Development of Tomato Fruits. Annals of Botany, 88(5), 869–877. https://doi.org/10.1006/anbo.2001.1524
4. Afreen, F. (2008). Temporary Immersion Bioreactor. In S. D. Gupta & Y. Ibaraki (Eds.), Plant Tissue Culture Engineering (pp. 187–201). Springer Netherlands. https://doi.org/10.1007/978-1-4020-3694-1_11
5. Alleweldt, G., & Radler, F. (1962). Interrelationship Between Photoperiodic Behavior of Grapes & Growth of Plant Tissue Cultures. Plant Physiology, 37(3), 376–379. https://www.ncbi.nlm.nih.gov/pmc/articles/PMC549796/
6. Álvarez, J. M., Pasian, C., Lal, R., López Núñez, R., & Fernández Martínez, M. (2017). Vermicompost and biochar as substitutes of growing media in ornamental-plant production. https://doi.org/10.13039/501100003329

7. Aremu, A. O., Bairu, M. W., Doležal, K., Finnie, J. F., & Van Staden, J. (2012). Topolins: A panacea to plant tissue culture challenges? Plant Cell, Tissue and Organ Culture (PCTOC), 108(1), 1–16. https://doi.org/10.1007/s11240-011-0007-7
8. Bairu, M. W., Aremu, A. O., & Van Staden, J. (2011). Somaclonal variation in plants: Causes and detection methods. Plant Growth Regulation, 63(2), 147–173. https://doi.org/10.1007/s10725-010-9554-x
9. Bedini, L., Lucchesini, M., Bertozzi, F., & Graifenberg, A. (2012). Plant tissue cultures from four Tuscan globe artichoke cultivars. Central European Journal of Biology, 7(4), 680–689. https://doi.org/10.2478/s11535-012-0064-x
10. Benson, E. E. (2000). Sepecial symposium: In vitro plant recalcitrance in vitro plant recalcitrance: An introduction. In Vitro Cellular & Developmental Biology - Plant, 36(3), 141–148. https://doi.org/10.1007/s11627-000-0029-z
11. Bhoite, H. A., & Palshikar, G. S. (2014). Plant Tissue Culture: A Review. World Journal of Pharmaceutical Sciences, 565–572. https://www.wjpsonline.com/index.php/wjps/article/view/plant-tissue-culture-review
12. Bhojwani, S. S., & Dantu, P. K. (2013a). General Requirements and Techniques. In S. S. Bhojwani & P. K. Dantu (Eds.), Plant Tissue Culture: An Introductory Text (pp. 11–25). Springer. https://doi.org/10.1007/978-81-322-1026-9_2
13. Bhojwani, S. S., & Dantu, P. K. (2013b). Production of Virus-Free Plants. In S. S. Bhojwani & P. K. Dantu (Eds.), Plant Tissue Culture: An Introductory Text (pp. 227–243). Springer. https://doi.org/10.1007/978-81-322-1026-9_16
14. Bi, Y.-M., Kenton, P., Mur, L., Darby, R., & Draper, J. (1995). Hydrogen peroxide does not function downstream of salicylic acid in the induction of PR protein expression. The Plant Journal, 8(2), 235–245. https://doi.org/10.1046/j.1365-313X.1995.08020235.x
15. Boosalis, M. G., & Scharen, A. L. (1959). Methods for microscopic detection of Aphanomyces eutiches and Rhizoctonia solani and for isolation of Rhizoctonia solani associated with plant debris. Phytopathology, 49(4). https://www.cabdirect.org/cabdirect/abstract/19591101753

16. Bozzo, G. G., Raghothama, K. G., & Plaxton, W. C. (2002). Purification and characterization of two secreted purple acid phosphatase isozymes from phosphate-starved tomato (Lycopersicon esculentum) cell cultures. European Journal of Biochemistry, 269(24), 6278–6286. https://doi.org/10.1046/j.1432-1033.2002.03347.x

17. Brown, P. H., & Shelp, B. J. (1997). Boron mobility in plants. Plant and Soil, 193(1), 85–101. https://doi.org/10.1023/A:1004211925160

18. Camargo, S., Rufato, L., Magro, M., & Kulkamp de Souza, A. L. (2019). Temporary immersion biorreators: Efficient technique for the propagation of the 'Pircinque' strawberry. Revista Brasileira de Fruticultura, 41. https://doi.org/10.1590/0100-29452019102

19. Carman, J. G. (1990). Embryogenic cells in plant tissue cultures: Occurrence and behavior. In Vitro Cellular &Developmental Biology, 26(8), 746–753. https://doi.org/10.1007/BF02623615

20. Cassells, A. C. (1991). Problems in tissue culture: Culture contamination. In P. C. Debergh & R. H. Zimmerman (Eds.), Micropropagation: Technology and Application (pp. 31–44). Springer Netherlands. https://doi.org/10.1007/978-94-009-2075-0_3

21. Cassells, A. C. (2012). Pathogen and Biological Contamination Management in Plant Tissue Culture: Phytopathogens, Vitro Pathogens, and Vitro Pests. In V. M. Loyola-Vargas & N. Ochoa-Alejo (Eds.), Plant Cell Culture Protocols (pp. 57–80). Humana Press. https://doi.org/10.1007/978-1-61779-818-4_6

22. Centeno, M. L., Rodríguez, A., Feito, I., & Fernández, B. (1996). Relationship between endogenous auxin and cytokinin levels and morphogenic responses inActinidia deliciosa tissue cultures. Plant Cell Reports, 16(1), 58–62. https://doi.org/10.1007/BF01275450

23. Chandra, S. (2012). Natural plant genetic engineer Agrobacterium rhizogenes: Role of T-DNA in plant secondary metabolism. Biotechnology Letters, 34(3), 407–415. https://doi.org/10.1007/s10529-011-0785-3

24. Chandra, S., Bandopadhyay, R., Kumar, V., & Chandra, R. (2010). Acclimatization of tissue cultured plantlets: From laboratory to land. Biotechnology Letters, 32(9), 1199–1205. https://doi.org/10.1007/s10529-010-0290-0

25. Chatterjee, J., Thakur, V., Nepomuceno, R., Coe, R. A., Dionora, J., Elmido-Mabilangan, A., Llave, A. D., Reyes, A. M. D., Monroy, A. N.,

Canicosa, I., Bandyopadhyay, A., Jena, K. K., Brar, D. S., & Quick, W. P. (2020). Natural Diversity in Stomatal Features of Cultivated and Wild Oryza Species. Rice, 13(1), 58. https://doi.org/10.1186/s12284-020-00417-0

26. Chen, L., Zhang, S., Beachy, R. N., & Fauquet, C. M. (1998). A protocol for consistent, large-scale production of fertile transgenic rice plants. Plant Cell Reports, 18(1), 25–31. https://doi.org/10.1007/s002990050526
27. Corredor-Prado, J. P., Schmidt, E. C., Guerra, M. P., Bouzon, Z. L., Dal Vesco, L. L., & Pescador, R. (2015). Histodifferentiation and ultrastructure of nodular cultures from seeds of Vriesea friburgensis Mez var. Paludosa (L.B. Smith) L.B. Smith and leaf explants of Vriesea reitzii Leme & A. Costa (Bromeliaceae). Journal of Microscopy and Ultrastructure, 3(4), 200–209. https://doi.org/10.1016/j.jmau.2015.04.001
28. Couselo, J. L., Corredoira, E., Vieitez, A. M., & Ballester, A. (2012). Plant Tissue Culture of Fast-Growing Trees for Phytoremediation Research. In V. M. Loyola-Vargas & N. Ochoa-Alejo (Eds.), Plant Cell Culture Protocols (pp. 247–263). Humana Press. https://doi.org/10.1007/978-1-61779-818-4_19
29. Cresswell, R. C., Fowler, M. W., Stafford, A., & Stepan-Sarkissian, G. (1989). Inputs and Outputs: Primary Substrates and Secondary Metabolism. In W. G. W. Kurz (Ed.), Primary and Secondary Metabolism of Plant Cell Cultures II (pp. 14–26). Springer. https://doi.org/10.1007/978-3-642-74551-5_2
30. Dagla, H. R. (2012). Plant tissue culture. Resonance, 17(8), 759–767. https://doi.org/10.1007/s12045-012-0086-8
31. Danby, S., Hampson, S. P., Joshi, S., Sigee, D. C., Epton, H. A. S., & Leifert, C. (1994). Activity of antibiotics produced by Bacillus subtilis and Bacillus pumilus against common fungal contaminants of plant tissue cultures. In P. J. Lumsden, J. R. Nicholas, & W. J. Davies (Eds.), Physiology, Growth and Development of Plants in Culture (pp. 404–408). Springer Netherlands. https://doi.org/10.1007/978-94-011-0790-7_45
32. Davey, M. R., Anthony, P., Power, J. B., & Lowe, K. C. (2005). Plant protoplasts: Status and biotechnological perspectives. Biotechnology Advances, 23(2), 131–171. https://doi.org/10.1016/j.biotechadv.2004.09.008

33. Davidonis, G. H., & Hamilton, R. H. (1983). Plant regeneration from callus tissue of Gossypium hirsutum L. Plant Science Letters, 32(1), 89–93. https://doi.org/10.1016/0304-4211(83)90102-5

34. Défago, G., Ammon, H. U., Cagán, L., Draeger, B., Greaves, M. P., Guntli, D., Hoeke, D., Klimes, L., Lawrie, J., Moënne-Loccoz, Y., Nicolet, B., Pfirter, H. A., Tabacchi, R., & Tóth, P. (2001). Towards the biocontrol of bindweeds with a mycoherbicide. BioControl, 46(2), 157–173. https://doi.org/10.1023/A:1011441816615

35. DiCosmo, F., & Misawa, M. (1995). Plant cell and tissue culture: Alternatives for metabolite production. Biotechnology Advances, 13(3), 425–453. https://doi.org/10.1016/0734-9750(95)02005-N

36. Dorai, M., Papadopoulos, A., & Gosselin, A. (2001). Influence of electric conductivity management on greenhouse tomato yield and fruit quality. Agronomie, 21(4), 367–383. https://doi.org/10.1051/agro:2001130

37. Ebert, A., & Taylor, H. F. (1990). Assessment of the changes of 2,4-dichlorophenoxyacetic acid concentrations in plant tissue culture media in the presence of activated charcoal. Plant Cell, Tissue and Organ Culture, 20(3), 165–172. https://doi.org/10.1007/BF00041877

38. Eibl, R., & Eibl, D. (2008). Design of bioreactors suitable for plant cell and tissue cultures. Phytochemistry Reviews, 7(3), 593–598. https://doi.org/10.1007/s11101-007-9083-z

39. Faisal, M., Alatar, A. A., Ahmad, N., Anis, M., & Hegazy, A. K. (2012). Assessment of Genetic Fidelity in Rauvolfia serpentina Plantlets Grown from Synthetic (Encapsulated) Seeds Following in Vitro Storage at 4 °C. Molecules, 17(5), Article 5. https://doi.org/10.3390/molecules17055050

40. Farbos, I., Oliveira, M., Negrutiu, I., & Mouras, A. (1997). Sex organ determination and differentiation in the dioecious plant Melandrium album (Silene latifolia): A cytological and histological analysis. Sexual Plant Reproduction, 10(3), 155–167. https://doi.org/10.1007/s004970050083

41. Fehér, A. (2006). Why Somatic Plant Cells Start to form Embryos? In A. Mujib & J. Šamaj (Eds.), Somatic Embryogenesis (pp. 85–101). Springer. https://doi.org/10.1007/7089_019

42. Filho, A. R., Vesco, L. L. D., Nodari, R. O., Lischka, R. W., Müller, C. V., & Guerra, M. P. (2005). Tissue culture for the conservation

and mass propagation of Vriesea reitzii Leme and Costa, a bromeliad threatened of extinction from the Brazilian Atlantic Forest. Biodiversity & Conservation, 14(8), 1799–1808. https://doi.org/10.1007/s10531-004-0700-5

43. Gamborg, O. L. (2002). Plant tissue culture. Biotechnology. Milestones. In Vitro Cellular & Developmental Biology - Plant, 38(2), 84–92. https://doi.org/10.1079/IVP2001281
44. Gaspar, T., Kevers, C., Penel, C., Greppin, H., Reid, D. M., & Thorpe, T. A. (1996). Plant hormones and plant growth regulators in plant tissue culture. In Vitro Cellular & Developmental Biology - Plant, 32(4), 272–289. https://doi.org/10.1007/BF02822700
45. Gautheret, R. J. (1955). The Nutrition of Plant Tissue Cultures. Annual Review of Plant Physiology, 6(1), 433–484. https://doi.org/10.1146/annurev.pp.06.060155.002245
46. Georgiev, M. I., Weber, J., & Maciuk, A. (2009). Bioprocessing of plant cell cultures for mass production of targeted compounds. Applied Microbiology and Biotechnology, 83(5), 809–823. https://doi.org/10.1007/s00253-009-2049-x
47. Germanà, M. A. (2011). Anther culture for haploid and doubled haploid production. Plant Cell, Tissue and Organ Culture (PCTOC), 104(3), 283–300. https://doi.org/10.1007/s11240-010-9852-z
48. Ghosh, A., Igamberdiev, A. U., & Debnath, S. C. (2018). Thidiazuron-induced somatic embryogenesis and changes of antioxidant properties in tissue cultures of half-high blueberry plants. Scientific Reports, 8(1), Article 1. https://doi.org/10.1038/s41598-018-35233-6
49. Gounaris, K., Barber, J., & Harwood, J. L. (1986). The thylakoid membranes of higher plant chloroplasts. Biochemical Journal, 237(2), 313–326. https://www.ncbi.nlm.nih.gov/pmc/articles/PMC1146990/
50. Graham, D. C. (1964). Taxonomy of the Soft Rot Coliform Bacteria. Annual Review of Phytopathology, 2(1), 13–42. https://doi.org/10.1146/annurev.py.02.090164.000305
51. Harms, K., & Wackernagel, W. (2008). The RecBCD and SbcCD DNases suppress homology-facilitated illegitimate recombination during natural transformation of Acinetobacter baylyi. Microbiology, 154(8), 2437–2445. https://doi.org/10.1099/mic.0.2008/018382-0
52. Hoffmann, N., King, S., Samuels, A. L., & McFarlane, H. E. (2021). Subcellular coordination of plant cell wall synthesis. Developmental Cell, 56(7), 933–948. https://doi.org/10.1016/j.devcel.2021.03.004

53. Horler, D. N. H., Barber, J., & Barringer, A. R. (1980). Effects of heavy metals on the absorbance and reflectance spectra of plants. International Journal of Remote Sensing, 1(2), 121–136. https://doi.org/10.1080/01431168008547550

54. Hosikian, A., Lim, S., Halim, R., & Danquah, M. K. (2010). Chlorophyll Extraction from Microalgae: A Review on the Process Engineering Aspects. International Journal of Chemical Engineering, 2010, e391632. https://doi.org/10.1155/2010/391632

55. Huang, C.-H., & Chu, C.-Y. (2008). Methods to improve the fertility of poinsettia. Scientia Horticulturae, 117(3), 271–274. https://doi.org/10.1016/j.scienta.2008.04.005

56. Huetteman, C. A., & Preece, J. E. (1993). Thidiazuron: A potent cytokinin for woody plant tissue culture. Plant Cell, Tissue and Organ Culture, 33(2), 105–119. https://doi.org/10.1007/BF01983223

57. Hüsemann, W., Fischer, K., Mittelbach, I., Hübner, S., Richter, G., & Barz, W. (1989). Photoautotrophic Plant Cell Cultures for Studies on Primary and Secondary Metabolism. In W. G. W. Kurz (Ed.), Primary and Secondary Metabolism of Plant Cell Cultures II (pp. 35–46). Springer. https://doi.org/10.1007/978-3-642-74551-5_4

58. Jack, E. M., Anatasova, S., & Verkleij, J. A. C. (2005). Callus induction and plant regeneration in the metallophyte Silene vulgaris (Caryophyllaceae). Plant Cell, Tissue and Organ Culture, 80(1), 25–31. https://doi.org/10.1007/s11240-004-5789-4

59. Jacobson, J. S., Weinstein, L. H., Mccune, D. C., & Hitchcock, A. E. (1966). The Accumulation of Fluorine by Plants. Journal of the Air Pollution Control Association, 16(8), 412–417. https://doi.org/10.1080/00022470.1966.10468494

60. Jay, V., Genestier, S., & Courduroux, J.-C. (1992). Bioreactor studies on the effect of dissolved oxygen concentrations on growth and differentiation of carrot (Daucus carota L.) cell cultures. Plant Cell Reports, 11(12), 605–608. https://doi.org/10.1007/BF00236382

61. Joko, T., Subandi, A., Kusumandari, N., Wibowo, A., & Priyatmojo, A. (2014). Activities of plant cell wall-degrading enzymes by bacterial soft rot of orchid. Archives of Phytopathology and Plant Protection, 47(10), 1239–1250. https://doi.org/10.1080/03235408.2013.838374

62. Kakkar, R. K., Nagar, P. K., Ahuja, P. S., & Rai, V. K. (2000). Polyamines and Plant Morphogenesis. Biologia Plantarum, 43(1), 1–11. https://doi.

org/10.1023/A:1026582308902

63. Kaplan, F., Kopka, J., Sung, D. Y., Zhao, W., Popp, M., Porat, R., & Guy, C. L. (2007). Transcript and metabolite profiling during cold acclimation of Arabidopsis reveals an intricate relationship of cold-regulated gene expression with modifications in metabolite content. The Plant Journal, 50(6), 967–981. https://doi.org/10.1111/j.1365-313X.2007.03100.x
64. Kasperbauer, M. J., & Collins, G. B. (1972). Reconstitution of Diploids from Leaf Tissue of Anther-derived Haploids in Tobacco1. Crop Science, 12(1), cropsci1972.0011183X001200010034x. https://doi.org/10.2135/cropsci1972.0011183X001200010034x
65. Khanday, I., & Sundaresan, V. (2021). Plant zygote development: Recent insights and applications to clonal seeds. Current Opinion in Plant Biology, 59, 101993. https://doi.org/10.1016/j.pbi.2020.101993
66. Kieran, P. M., MacLoughlin, P. F., & Malone, D. M. (1997). Plant cell suspension cultures: Some engineering considerations. Journal of Biotechnology, 59(1), 39–52. https://doi.org/10.1016/S0168-1656(97)00163-6
67. Kim, J. Y., Adhikari, P. B., Ahn, C. H., Kim, D. H., Chang Kim, Y., Han, J. Y., Kondeti, S., & Choi, Y. E. (2019). High frequency somatic embryogenesis and plant regeneration of interspecific ginseng hybrid between Panax ginseng and Panax quinquefolius. Journal of Ginseng Research, 43(1), 38–48. https://doi.org/10.1016/j.jgr.2017.08.002
68. Kim, M. K., Ryu, J. S., Lee, Y. H., & Yun, H. D. (2007). First Report of Pantoea sp. Induced Soft Rot Disease of Pleurotus eryngii in Korea. Plant Disease, 91(1), 109–109. https://doi.org/10.1094/PD-91-0109A
69. Kumar, A., Taylor, M., Altabella, T., & Tiburcio, A. F. (1997). Recent advances in polyamine research. Trends in Plant Science, 2(4), 124–130. https://doi.org/10.1016/S1360-1385(97)01013-3
70. Kumar, P. P., & Loh, C. S. (2012). 9—Plant tissue culture for biotechnology. In A. Altman & P. M. Hasegawa (Eds.), Plant Biotechnology and Agriculture (pp. 131–138). Academic Press. https://doi.org/10.1016/B978-0-12-381466-1.00009-2
71. Kutschera, U., & Ray, P. M. (2022). Forever young: Stem cell and plant regeneration one century after Haberlandt 1921. Protoplasma, 259(1), 3–18. https://doi.org/10.1007/s00709-021-01683-5
72. Lakshmanan, K. K., & Ambegaokar, K. B. (1984). Polyembryony. In B.

M. Johri (Ed.), Embryology of Angiosperms (pp. 445–474). Springer. https://doi.org/10.1007/978-3-642-69302-1_9

73. Larbi, A., Morales, F., Abadía, A., Gogorcena, Y., Lucena, J. J., & Abadía, J. (2002). Effects of Cd and Pb in sugar beet plants grown in nutrient solution: Induced Fe deficiency and growth inhibition. Functional Plant Biology, 29(12), 1453–1464. https://doi.org/10.1071/fp02090
74. Leifert, C., & Cassells, A. C. (2001). Microbial hazards in plant tissue and cell cultures. In Vitro Cellular & Developmental Biology - Plant, 37(2), 133–138. https://doi.org/10.1007/s11627-001-0025-y
75. Leifert, C., Murphy, K. P., & Lumsden, P. J. (1995). Mineral and Carbohydrate Nutrition of Plant Cell and Tissue Cultures. Critical Reviews in Plant Sciences, 14(2), 83–109. https://doi.org/10.1080/07352689509701923
76. Leifert, C., Ritchie, J. Y., & Waites, W. M. (1991). Contaminants of plant-tissue and cell cultures. World Journal of Microbiology and Biotechnology, 7(4), 452–469. https://doi.org/10.1007/BF00303371
77. Leifert, C., & Waites, W. m. (1992). Bacterial growth in plant tissue culture media. Journal of Applied Bacteriology, 72(6), 460–466. https://doi.org/10.1111/j.1365-2672.1992.tb01859.x
78. Li, F., Cui, X., Feng, Z., Du, X., & Zhu, J. (2012). The effect of 2,4-D and kinetin on dedifferentiation of petiole cells in Arabidopsis thaliana. Biologia Plantarum, 56(1), 121–125. https://doi.org/10.1007/s10535-012-0026-1
79. Liu, G., Li, C., Yu, H., Tao, P., Yuan, L., Ye, J., Chen, W., Wang, Y., Ge, P., Zhang, J., Zhou, G., Zheng, W., Ye, Z., & Zhang, Y. (2020). GREEN STRIPE, encoding methylated TOMATO AGAMOUS-LIKE 1, regulates chloroplast development and Chl synthesis in fruit. New Phytologist, 228(1), 302–317. https://doi.org/10.1111/nph.16705
80. Liu, M., Rehman, S., Tang, X., Gu, K., Fan, Q., Chen, D., & Ma, W. (2019). Methodologies for Improving HDR Efficiency. Frontiers in Genetics, 9. https://www.frontiersin.org/articles/10.3389/fgene.2018.00691
81. Lonigro, A., Montemurro, N., & Laera, G. (2017). Effects of residual disinfectant on soil and lettuce crop irrigated with chlorinated water. Science of The Total Environment, 584–585, 595–602. https://doi.org/10.1016/j.scitotenv.2017.01.083

82. Lowe, K. C., Davey, M. R., & Power, J. B. (1998). Perfluorochemicals: Their applications and benefits to cell culture. Trends in Biotechnology, 16(6), 272–277. https://doi.org/10.1016/S0167-7799(98)01205-0
83. Loyola-Vargas, V. M., & Vázquez-Flota, F. (2006). An Introduction to Plant Cell Culture. In V. M. Loyola-Vargas & F. Vázquez-Flota (Eds.), Plant Cell Culture Protocols (pp. 3–8). Humana Press. https://doi.org/10.1385/1-59259-959-1:003
84. Maene, L., & Debergh, P. (1985). Liquid medium additions to established tissue cultures to improve elongation and rooting in vivo. Plant Cell, Tissue and Organ Culture, 5(1), 23–33. https://doi.org/10.1007/BF00033566
85. Malik, S., Hossein Mirjalili, M., Fett-Neto, A. G., Mazzafera, P., & Bonfill, M. (2013). Living between two worlds: Two-phase culture systems for producing plant secondary metabolites. Critical Reviews in Biotechnology, 33(1), 1–22. https://doi.org/10.3109/07388551.2012.659173
86. McKnight, T. D., Lillis, M. T., & Simpson, R. B. (1987). Segregation of genes transferred to one plant cell from two separate Agrobacterium strains. Plant Molecular Biology, 8(6), 439–445. https://doi.org/10.1007/BF00017989
87. Meï, C., Michaud, M., Cussac, M., Albrieux, C., Gros, V., Maréchal, E., Block, M. A., Jouhet, J., & Rébeillé, F. (2015). Levels of polyunsaturated fatty acids correlate with growth rate in plant cell cultures. Scientific Reports, 5(1), Article 1. https://doi.org/10.1038/srep15207
88. Minz, A., & Sharon, A. (2010). Electroporation and Agrobacterium-Mediated Spore Transformation. In A. Sharon (Ed.), Molecular and Cell Biology Methods for Fungi (pp. 21–32). Humana Press. https://doi.org/10.1007/978-1-60761-611-5_2
89. Mitrofanova, I., Tsyupka, V., & Jain, S. M. (2022). Chapter 8—Morpho-anatomical characterization of in vitro regenerated plants. In A. Chandra Rai, A. Kumar, A. Modi, & M. Singh (Eds.), Advances in Plant Tissue Culture (pp. 175–204). Academic Press. https://doi.org/10.1016/B978-0-323-90795-8.00018-7
90. Mobin, M., Wu, C.-H., Tewari, R. K., & Paek, K.-Y. (2015). Studies on the glyphosate-induced amino acid starvation and addition of precursors on caffeic acid accumulation and profiles in adventitious roots of Echinacea purpurea (L.) Moench. Plant Cell, Tissue and Organ Culture (PCTOC), 120(1), 291–301. https://doi.org/10.1007/s11240-

014-0606-1

91. Mukhopadhyay, A., & Bhojwani, S. S. (1978). Shoot-Bud Differentiation in Tissue Cultures of Leguminous Plants. Zeitschrift Für Pflanzenphysiologie, 88(3), 263–268. https://doi.org/10.1016/S0044-328X(78)80248-7
92. Murashige, T. (1974). Plant Propagation Through Tissue Cultures. Annual Review of Plant Physiology, 25(1), 135–166. https://doi.org/10.1146/annurev.pp.25.060174.001031
93. Neumann, K.-H., Kumar, A., & Imani, J. (2020). Callus Cultures. In K.-H. Neumann, A. Kumar, & J. Imani (Eds.), Plant Cell and Tissue Culture – A Tool in Biotechnology: Basics and Application (pp. 25–59). Springer International Publishing. https://doi.org/10.1007/978-3-030-49098-0_3
94. Nieves–Brun, C. (1985). Infection of roots of Dieffenbachia maculata by the foliar blight and soft rot pathogen, Erwinia chrysanthemi. Plant Pathology, 34(1), 139–145. https://doi.org/10.1111/j.1365-3059.1985.tb02770.x
95. Norman, D. J., Chen, J., Yuen, J. M. F., Mangravita-Novo, A., Byrne, D., & Walsh, L. (2006). Control of Bacterial Wilt of Geranium with Phosphorous Acid. Plant Disease, 90(6), 798–802. https://doi.org/10.1094/PD-90-0798
96. Norstog, K. (1973). New synthetic medium for the culture of premature barley embryos. In Vitro, 8(4), 307–308. https://doi.org/10.1007/BF02615911
97. Oliver, M. J., Farrant, J. M., Hilhorst, H. W. M., Mundree, S., Williams, B., & Bewley, J. D. (2020). Desiccation Tolerance: Avoiding Cellular Damage During Drying and Rehydration. Annual Review of Plant Biology, 71(1), 435–460. https://doi.org/10.1146/annurev-arplant-071219-105542
98. Orlikowska, T., Nowak, K., & Reed, B. (2017). Bacteria in the plant tissue culture environment. Plant Cell, Tissue and Organ Culture (PCTOC), 128(3), 487–508. https://doi.org/10.1007/s11240-016-1144-9
99. Pachota, K. A., & Orłowska, R. (2022). Effect of copper and silver ions on sequence and DNA methylation changes in triticale regenerants gained via somatic embryogenesis. Journal of Applied Genetics, 63(4), 663–675. https://doi.org/10.1007/s13353-022-00717-9

100. Panda, A. K., Mishra, S., Bisaria, V. S., & Bhojwani, S. S. (1989). Plant cell reactors—A perspective. Enzyme and Microbial Technology, 11(7), 386–397. https://doi.org/10.1016/0141-0229(89)90132-4
101. Pant, B., & Thapa, D. (2012). In vitro mass propagation of an epiphytic orchid, Dendrobium primulinum Lindl. Through shoot tip culture. African Journal of Biotechnology, 11(42), Article 42. https://doi.org/10.4314/ajb.v11i42
102. Pépin, M.-F., Smith, M. A. L., & Reid, J. F. (1999). Application of imaging tools to plant cell culture: Relationship between plant cell aggregation and flavonoid production. In Vitro Cellular & Developmental Biology - Plant, 35(4), 290–295. https://doi.org/10.1007/s11627-999-0036-7
103. Phillips, G. C., & Garda, M. (2019). Plant tissue culture media and practices: An overview. In Vitro Cellular & Developmental Biology - Plant, 55(3), 242–257. https://doi.org/10.1007/s11627-019-09983-5
104. Poorinmohammad, N., & Kerkhoven, E. J. (2021). Systems-level approaches for understanding and engineering of the oleaginous cell factory Yarrowia lipolytica. Biotechnology and Bioengineering, 118(10), 3640–3654. https://doi.org/10.1002/bit.27859
105. Prammanee, S., Thumjamras, S., Chiemsombat, P., & Pipattanawong, N. (2011). Efficient shoot regeneration from direct apical meristem tissue to produce virus-free purple passion fruit plants. Crop Protection, 30(11), 1425–1429. https://doi.org/10.1016/j.cropro.2011.07.008
106. Quintero, C., & Bowers, M. D. (2011). Plant Induced Defenses Depend More on Plant Age than Previous History of Damage: Implications for Plant-Herbivore Interactions. Journal of Chemical Ecology, 37(9), 992. https://doi.org/10.1007/s10886-011-0007-4
107. Ramos, A. C., Dobbss, L. B., Santos, L. A., Fernandes, M. S., Olivares, F. L., Aguiar, N. O., & Canellas, L. P. (2015). Humic matter elicits proton and calcium fluxes and signaling dependent on Ca2+-dependent protein kinase (CDPK) at early stages of lateral plant root development. Chemical and Biological Technologies in Agriculture, 2(1), 3. https://doi.org/10.1186/s40538-014-0030-0
108. Rathore, K. S., & Goldsworthy, A. (1985). Electrical Control of Shoot Regeneration in Plant Tissue Cultures. Bio/Technology, 3(12), Article 12. https://doi.org/10.1038/nbt1285-1107
109. Reddy, K. R., & De Busk, W. F. (1985). Nutrient Removal Potential of Selected Aquatic Macrophytes. Journal of Environmental Quality, 14(4), 459–462. https://doi.org/10.2134/jeq1985.00472425001400040001x

110. Rice, A. V., & Currah, R. S. (2006). Oidiodendron maius: Saprobe in Sphagnum Peat, Mutualist in Ericaceous Roots? In B. J. E. Schulz, C. J. C. Boyle, & T. N. Sieber (Eds.), Microbial Root Endophytes (pp. 227–246). Springer. https://doi.org/10.1007/3-540-33526-9_13
111. Roberts, J. A., & Hooley, R. (1988). Biosynthesis and Metabolism—More than Making and Breaking. In J. A. Roberts & R. Hooley (Eds.), Plant Growth Regulators (pp. 4–33). Springer US. https://doi.org/10.1007/978-1-4615-7592-4_2
112. Rocha, V., & Ting, I. P. (1971). Malate dehydrogenases of leaf tissue from Spinacia oleracea: Properties of three isoenzymes. Archives of Biochemistry and Biophysics, 147(1), 114–122. https://doi.org/10.1016/0003-9861(71)90316-X
113. Rowe, W. J. (1986). New Technologies in Plant Tissue Culture. In R. H. Zimmerman, R. J. Griesbach, F. A. Hammerschlag, & R. H. Lawson (Eds.), Tissue culture as a plant production system for horticultural crops: Conference on Tissue Culture as a Plant Production System for Horticultural Crops, Beltsville, MD, October 20–23, 1985 (pp. 35–51). Springer Netherlands. https://doi.org/10.1007/978-94-009-4444-2_3
114. Sachs, T., & Thimann, K. V. (1967). The Role of Auxins and Cytokinins in the Release of Buds from Dominance. American Journal of Botany, 54(1), 136–144. https://doi.org/10.1002/j.1537-2197.1967.tb06901.x
115. Sandhir, R., Garg, S. K., & Modi, D. R. (1999). GENETICS OF MICROORGANISMS | Fungi. In R. K. Robinson (Ed.), Encyclopedia of Food Microbiology (pp. 921–929). Elsevier. https://doi.org/10.1006/rwfm.1999.0735
116. Satdive, R. K., Fulzele, D. P., & Eapen, S. (2007). Enhanced production of azadirachtin by hairy root cultures of Azadirachta indica A. Juss by elicitation and media optimization. Journal of Biotechnology, 128(2), 281–289. https://doi.org/10.1016/j.jbiotec.2006.10.009
117. Schenke, D., Utami, H. P., Zhou, Z., Gallegos, M.-T., & Cai, D. (2019). Suppression of UV-B stress induced flavonoids by biotic stress: Is there reciprocal crosstalk? Plant Physiology and Biochemistry, 134, 53–63. https://doi.org/10.1016/j.plaphy.2018.06.026
118. Schimel, D. S., Braswell, B. H., McKeown, R., Ojima, D. S., Parton, W. J., & Pulliam, W. (1996). Climate and nitrogen controls on the geography and timescales of terrestrial biogeochemical cycling. Global Biogeochemical Cycles, 10(4), 677–692. https://doi.org/10.1029/96GB01524

119. Scoccianti, V., Bucchini, A. E., Iacobucci, M., Ruiz, K. B., & Biondi, S. (2016). Oxidative stress and antioxidant responses to increasing concentrations of trivalent chromium in the Andean crop species Chenopodium quinoa Willd. Ecotoxicology and Environmental Safety, 133, 25–35. https://doi.org/10.1016/j.ecoenv.2016.06.036
120. Shahzad, A., Sharma, S., Parveen, S., Saeed, T., Shaheen, A., Akhtar, R., Yadav, V., Upadhyay, A., & Ahmad, Z. (2017). Historical Perspective and Basic Principles of Plant Tissue Culture. In M. Z. Abdin, U. Kiran, Kamaluddin, & A. Ali (Eds.), Plant Biotechnology: Principles and Applications (pp. 1–36). Springer. https://doi.org/10.1007/978-981-10-2961-5_1
121. Sihachakr, D., & Ducreux, G. (1987). Cultural behavior of protoplasts from different organs of eggplant (solanum melongena L.), and plant regeneration. Plant Cell, Tissue and Organ Culture, 11(3), 179–188. https://doi.org/10.1007/BF00040424
122. Singh, V., Chauhan, N. S., Singh, M., Idris, A., Madanala, R., Pande, V., & Mohanty, C. S. (2014). Establishment of an efficient and rapid method of multiple shoot regeneration and a comparative phenolics profile in in vitro and greenhouse-grown plants of psophocarpus tetragonolobus (L.) DC. Plant Signaling & Behavior, 9(10), e970443. https://doi.org/10.4161/15592316.2014.970443
123. Skirvin, R. M., Chu, M. C., Mann, M. L., Young, H., Sullivan, J., & Fermanian, T. (1986). Stability of tissue culture medium pH as a function of autoclaving, time, and cultured plant material. Plant Cell Reports, 5(4), 292–294. https://doi.org/10.1007/BF00269825
124. Soni, R. A., Sudhakar, K., & Rana, R. S. (2017). Spirulina – From growth to nutritional product: A review. Trends in Food Science & Technology, 69, 157–171. https://doi.org/10.1016/j.tifs.2017.09.010
125. Sowley, E. N. K., Dewey, F. M., & Shaw, M. W. (2010). Persistent, symptomless, systemic, and seed-borne infection of lettuce by Botrytis cinerea. European Journal of Plant Pathology, 126(1), 61–71. https://doi.org/10.1007/s10658-009-9524-1
126. Sugiyama, M. (1999). Organogenesis in vitro. Current Opinion in Plant Biology, 2(1), 61–64. https://doi.org/10.1016/S1369-5266(99)80012-0
127. Sugiyama, M. (2015). Historical review of research on plant cell dedifferentiation. Journal of Plant Research, 128(3), 349–359. https://doi.org/10.1007/s10265-015-0706-y

128. Tamas, I. A. (1995). Hormonal Regulation of Apical Dominance. In P. J. Davies (Ed.), Plant Hormones: Physiology, Biochemistry and Molecular Biology (pp. 572–597). Springer Netherlands. https://doi.org/10.1007/978-94-011-0473-9_27
129. Tanaka-Ueguchi, M., Itoh, H., Oyama, N., Koshioka, M., & Matsuoka, M. (1998). Over-expression of a tobacco homeobox gene, NTH15 , decreases the expression of a gibberellin biosynthetic gene encoding GA 20-oxidase. The Plant Journal, 15(3), 391–400. https://doi.org/10.1046/j.1365-313X.1998.00217.x
130. Teixeira da Silva, J. A., Dobránszki, J., Winarto, B., & Zeng, S. (2015). Anthurium in vitro: A review. Scientia Horticulturae, 186, 266–298. https://doi.org/10.1016/j.scienta.2014.11.024
131. Teixeira, J. B., Söndahl, M. R., Nakamura, T., & Kirby, E. G. (1995). Establishment of oil palm cell suspensions and plant regeneration. Plant Cell, Tissue and Organ Culture, 40(2), 105–111. https://doi.org/10.1007/BF00037662
132. Tepfer, D. (1983). The Potential uses of Agrobacterium Rhizogenes in the Genetic Engineering of Higher Plants: Nature Got There First. In P. F. Lurquin & A. Kleinhofs (Eds.), Genetic Engineering in Eukaryotes (pp. 153–164). Springer US. https://doi.org/10.1007/978-1-4684-4493-3_17
133. Thorpe, T. A. (2007). History of plant tissue culture. Molecular Biotechnology, 37(2), 169–180. https://doi.org/10.1007/s12033-007-0031-3
134. Travella, S., Ross, S. M., Harden, J., Everett, C., Snape, J. W., & Harwood, W. A. (2005). A comparison of transgenic barley lines produced by particle bombardment and Agrobacterium-mediated techniques. Plant Cell Reports, 23(12), 780–789. https://doi.org/10.1007/s00299-004-0892-x
135. Tsai, C.-J., & Saunders, J. W. (1999). Evaluation of sole nitrogen sources for shoot and leaf disc cultures of sugarbeet. Plant Cell, Tissue and Organ Culture, 59(1), 47–56. https://doi.org/10.1023/A:1006344514809
136. Twaij, B. M., Jazar, Z. H., & Hasan, M. N. (2020). Trends in the Use of Tissue Culture, Applications and Future Aspects. International Journal of Plant Biology, 11(1), Article 1. https://doi.org/10.4081/pb.2020.8385
137. Utami, E. S. W., & Hariyanto, S. (2019). In Vitro Seed Germination and Seedling Development of a Rare Indonesian Native Orchid

Phalaenopsis amboinensis J.J.Sm. Scientifica, 2019, e8105138. https://doi.org/10.1155/2019/8105138

138. Vasil, I. K., & Vasil, V. (1972). Totipotency and embryogenesis in plant cell and tissue cultures. In Vitro, 8(3), 117–125. https://doi.org/10.1007/BF02619487

139. Verpoorte, R., Heijden, R. van der, Schripsema, J., Hoge, J. H. C., & Hoopen, H. J. G. T. (2004, July 1). Plant Cell Biotechnology for the Production of Alkaloids: Present Status and Prospects (world). ACS Publications; American Chemical Society. https://doi.org/10.1021/np50092a003

140. Wagner, F., & Vogelmann, H. (1977). Cultivation of Plant Tissue Cultures in Bioreactors and Formation of Secondary Metabolites. In W. Barz, E. Reinhard, & M. H. Zenk (Eds.), Plant Tissue Culture and Its Bio-technological Application (pp. 245–252). Springer. https://doi.org/10.1007/978-3-642-66646-9_20

141. Walker, K. A., Wendeln, M. L., & Jaworski, E. G. (1979). Organogenesis in callus tissue of Medicago sativa. The temporal separation of induction processes from differentiation processes. Plant Science Letters, 16(1), 23–30. https://doi.org/10.1016/0304-4211(79)90004-X

142. Wang, P. J., & Hu, C. Y. (1980). Regeneration of virus-free plants through in vitro culture. In A. Fiechter (Ed.), Advances in Biomedical Engineering, Volume 18 (pp. 61–99). Springer. https://doi.org/10.1007/3-540-09936-0_3

143. Weathers, P. J., Towler, M. J., & Xu, J. (2010). Bench to batch: Advances in plant cell culture for producing useful products. Applied Microbiology and Biotechnology, 85(5), 1339–1351. https://doi.org/10.1007/s00253-009-2354-4

144. White, F. F., Taylor, B. H., Huffman, G. A., Gordon, M. P., & Nester, E. W. (1985). Molecular and genetic analysis of the transferred DNA regions of the root-inducing plasmid of Agrobacterium rhizogenes. Journal of Bacteriology, 164(1), 33–44. https://doi.org/10.1128/jb.164.1.33-44.1985

145. Widholm, J. M. (1989). Initiation and Characterization of Photoautotrophic Suspension Cultures. In W. G. W. Kurz (Ed.), Primary and Secondary Metabolism of Plant Cell Cultures II (pp. 3–13). Springer. https://doi.org/10.1007/978-3-642-74551-5_1

146. Widholm, J. M. (1992). Properties and Uses of Photoautotrophic Plant

Cell Cultures. In K. W. Jeon & M. Friedlander (Eds.), International Review of Cytology (Vol. 132, pp. 109–175). Academic Press. https://doi.org/10.1016/S0074-7696(08)62455-9

147. Williamson, B., Duncan, G. H., Harrison, J. G., Harding, L. A., Elad, Y., & Zimand, G. (1995). Effect of humidity on infection of rose petals by dry-inoculated conidia of Botrytis cinerea. Mycological Research, 99(11), 1303–1310. https://doi.org/10.1016/S0953-7562(09)81212-4
148. Xia, Y., Cao, Y., Ren, Y., Ling, A., Du, K., Li, Y., Yang, J., & Kang, X. (2023). Effect of a suitable treatment period on the genetic transformation efficiency of the plant leaf disc method. Plant Methods, 19(1), 15. https://doi.org/10.1186/s13007-023-00994-3
149. Xu, J., Ge, X., & Dolan, M. C. (2011). Towards high-yield production of pharmaceutical proteins with plant cell suspension cultures. Biotechnology Advances, 29(3), 278–299. https://doi.org/10.1016/j.biotechadv.2011.01.002
150. Yamaguchi, S. (2008). Gibberellin Metabolism and its Regulation. Annual Review of Plant Biology, 59(1), 225–251. https://doi.org/10.1146/annurev.arplant.59.032607.092804
151. Yang, C., Hua, Q., & Shimizu, K. (2000). Energetics and carbon metabolism during growth of microalgal cells under photoautotrophic, mixotrophic and cyclic light-autotrophic/dark-heterotrophic conditions. Biochemical Engineering Journal, 6(2), 87–102. https://doi.org/10.1016/S1369-703X(00)00080-2
152. Yeung, E. C. (1999). The use of histology in the study of plant tissue culture systems—Some practical comments. In Vitro Cellular & Developmental Biology - Plant, 35(2), 137–143. https://doi.org/10.1007/s11627-999-0023-z
153. Yue, W., Ming, Q., Lin, B., Rahman, K., Zheng, C.-J., Han, T., & Qin, L. (2016). Medicinal plant cell suspension cultures: Pharmaceutical applications and high-yielding strategies for the desired secondary metabolites. Critical Reviews in Biotechnology, 36(2), 215–232. https://doi.org/10.3109/07388551.2014.923986
154. Zaid, A., El-Korchi, B., & Visser, H. J. (2011). Commercial Date Palm Tissue Culture Procedures and Facility Establishment. In S. M. Jain, J. M. Al-Khayri, & D. V. Johnson (Eds.), Date Palm Biotechnology (pp. 137–180). Springer Netherlands. https://doi.org/10.1007/978-94-007-1318-5_8

155. Zimmerman, R. H. (1985). Application of Tissue Culture Propagation to Woody Plants. In R. R. Henke, K. W. Hughes, M. J. Constantin, A. Hollaender, & C. M. Wilson (Eds.), Tissue Culture in Forestry and Agriculture (pp. 165–177). Springer US. https://doi.org/10.1007/978-1-4899-0378-5_13

156. Zobayed, S. M. A., Afreen, F., Kubota, C., & Kozai, T. (2000). Water Control and Survival of Ipomoea batatas Grown Photoautotrophically under Forced Ventilation and Photomixotrophically under Natural Ventilation. Annals of Botany, 86(3), 603–610. https://doi.org/10.1006/anbo.2000.1225

157. Zuniga-Soto, E., Mullins, E., & Dedicova, B. (2015). Ensifer-mediated transformation: An efficient non-Agrobacterium protocol for the genetic modification of rice. SpringerPlus, 4(1), 600. https://doi.org/10.1186/s40064-015-1369-9

INDEX

R

S

T